The supposedly beneficial hand of
to be a brutal boot, stamping on
alike. But there is no going back to the discredited
style socialism. Saral Sarkar's new study shows that there is a genuine third way, one which embraces an awareness of ecological limits within a framework of social solidarity, a vision truly worthy of the new millennium. *Sandy Irvine, Associate Editor, The Ecologist*

A very important book for those concerned with whether or not it is possible to fuse the radical ecology and socialist movements. Sarkar believes it is possible, providing socialism is prepared to redefine itself and learn 'the ecological lesson' from the radical ecology movement. I highly recommend this book for its ecological critique of all forms of socialism, its critique of green politics, and its insightful examination of traditional cultures and what can be learnt from them. *David Orton, Coordinator, Green Web environmental research group, Nova Scotia, Canada*

Saral Sarkar's critical analysis of the relationship between socialism and the ecological problems created by rapid industrialisation is highly instructive, particularly since it is written by a Third World scholar seriously concerned with development that is really sustainable for all peoples. He show with much revealing evidence that 'socialist' economies, with the ecological havoc they created, were not really socialist and that eco-socialism could (and should) be a far better alternative to the prevailing capitalist models, including eco-capitalism. A radical view which will help serious and badly needed questioning of the economic and ecological paradigms. *Professor Gerrit Huizer, Catholic University of Nijmegen*

One of the first of what one hopes will be many studies, theoretical and otherwise, of what it takes to combine and sublate socialism and ecology. The author offers an austere brand of ecosocialism, which readers may reject, but which ecosocialists need to take seriously. *James O'Connor, Founding Editor, Capitalism, Nature, Socialism*

Sarkar argues forcefully for a socialism that accepts ecological limits to growth. It is a case that socialists should heed if they take the principle of international equality seriously. Sarkar draws lessons from the failures of capitalist and socialist industrialism as well as ecological lessons from the South to make a convincing case for eco-socialism. *Mary Mellor, University of Northumbria*

About the author

Saral Sarkar was born in 1936 in West Bengal. After graduating from the University of Calcutta, he studied German language and literature for five years at the Goethe Institute, where from 1966 to 1981 he taught in Hyderabad. Since 1982 he has been living in Cologne, Germany, where he has been active in the Green movement, including for a time as a member of the Green Party. He has taken part in many debates on Green and alternative politics over the years and has published widely in journals in India, various European countries and the United States. He is the author of the authoritative historical work *Green Alternative Politics in West Germany* (2 vols) (United Nations University Press, 1993 and 1994).

Eco-Socialism or Eco-Capitalism?

A Critical Analysis of Humanity's Fundamental Choices

Saral Sarkar

If the protection of Nature thesis is to be helpful - and hence effective - in any additional way it must take into account the problems of practical action

1. practice
2. the scales of political efficacy to which it is attached.

practice/ history/ science/ bay/ Nature/ culture/ pol: sci/ ethics/ economics

Zed Books
LONDON AND NEW YORK

Eco-Socialism or Eco-Capitalism? was first published by
Zed Books Ltd, 7 Cynthia Street, London N1 9JF, UK and
Room 400, 175 Fifth Avenue, New York, NY 10010, USA
in 1999.

Distributed exclusively in the USA by St Martin's Press, Inc.,
175 Fifth Avenue, New York, NY 10010, USA.

Cover designed by Andrew Corbett
Set in Monotype Dante by Ewan Smith
Printed and bound in Malaysia

A catalogue record for this book is available from the British Library

US CIP has been applied for from the Library of Congress

ISBN 1 85649 599 x cased
ISBN 1 85649 600 7 limp

Contents

Acknowledgements

I am intellectually indebted to many thinkers, scholars, and researchers, whose works have contributed to the development of my own thoughts. From the works of many of them I have also quoted extensively in order to present evidence in support of my theses. They are so many in number that I cannot mention them here. That is also not necessary. The reader will find their names in the body of the book and the bibliography.

Professor Gerrit Huizer, Pieter Jansan, Wouter van Eck, Bob Tatam, David Orton, Ralph Smith (who also copy-edited the book), Robert Molteno and two external readers of the publishers have read the manuscript and given valuable comments and suggestions. I thank them sincerely for the time and energy they have spent for this purpose.

For direct and indirect, concrete and moral support they have given, I also thank sincerely Hermine Karas, Fritz Karas, Rosalind Tatam, Govind Kelkar and, last but not least, my wife Maria Mies.

CHAPTER I

Introduction

THE NEED FOR AN OVERALL ANALYSIS AND AN OVERALL PERSPECTIVE

The idea to write this book arose in early 1990. The Berlin Wall fell in November 1989, and thereafter all the 'socialist' regimes of Eastern Europe fell like dominoes. Socialists, in any sense of the term, were deeply shocked, or at least deeply disappointed. Not all of them had considered these regimes to have been really socialist. A new term had, therefore, been invented to characterise such societies, namely 'actually existing socialism'. But many had cherished the hope that these societies could at least develop towards the ideal of socialism.

Greens and environmentalists experienced a similar disappointment, although it was less of a shock. Since 1972, when *Limits to Growth*, the first report to the Club of Rome, was published, nothing had been undertaken anywhere in the world by any government to stop the spiral of growth. On the contrary, all governments continued to push it upwards. By the early 1980s, the ecology movement, where it existed, subsided – all sections of it.

In the second half of the 1980s, many Greens and environmentalists, at least in Germany, experienced a further disappointment. In 1985, the Green Party became the coalition partner of the Social Democrats in the provincial government of Hessen at the cost of giving up all its radical positions. In the course of the following years, this became the general political line of the whole party (for details see Sarkar 1994). Large sections of the ecology movement became, like the Green Party, adherents of 'pragmatism', and their practice consisted mainly of lobbying (for details see Sarkar 1993).

I experienced both disappointments personally, because I have for long been a socialist, and I was both an active member of the Green Party of West Germany and an activist of the ecology and peace movements.

The (alleged or apprehended) failure of socialism[1] had been a topic of discussion and debate even before 1989. But such debates were connected with the hope that the 'socialist' societies would succeed in reforming themselves and progress again towards real socialism. Or it was hoped that the Chinese or North Korean or Albanian communists would succeed, following a different path, in remaining or becoming really socialist societies. But since 1989, there has been no hope. As regards China today, what obtains there is capitalism or a mixed economy under the rule of a so-called communist party. For a few years after 1989, Cuba seemed to remain a bastion of 'socialism'. But now it seems that Cuba has decided to follow the Chinese path. Many leftists have reached an accommodation with capitalism.

A similar development took place in the ecology movement in Germany (I take this example because the movement there was very strong). Whereas in the second half of the 1970s and the early 1980s radical ecologists dominated the discourse and projected an eco-radical society or economy opposed, and alternative, to industrial society, by the middle of the 1980s, ideas of ecologically restructuring or modernising industrial society became dominant. This was strengthened through the idea of sustainable development. And since the end of 'socialism', many leading environmentalists are visualising eco-capitalism as their goal.

In the Green Party of West Germany, a parallel development took place, but with the special feature that in it, for several years, leftists dominated. They combined the idea of an ecological restructuring of industrial society with their traditional leftist politics. But today, the Green Party's ideal is just eco-capitalism with a slightly stronger social component (see Fischer 1989; Die Grünen 1986).

After 1989, in discussions on socialism, the cause(s) of the failure of 'socialist' societies was the main topic. Some socialists dismissed the question by saying that there has never been a socialist society, and therefore no discussion of the failure of socialism could take place. What failed in Eastern Europe, they said, was something else.

This, I think, is mere polemics. It is of no use. Of course, one may be a purist and compare the former 'socialist' regimes with strict criteria, and then conclude that they were not socialist. Of course, there were so many similarities between the capitalist West and 'socialist' Eastern Europe that many even talked of a convergence of the systems. I myself consider them to be two variants of the same thing, namely industrial society. But the two variants also had differences, which is why they were two variants. I think it is very important for socialists to search for the causes of failure of the 'socialist' regimes. I, like many others, have solved the terminological problem by using 'socialism' and 'socialist' (within inverted commas) when writing about these regimes and their system. I have not used the inverted commas in a pejorative sense, but simply to indicate that I mean this particular system. For all other conceptions of socialism, I have used the quite vague term without inverted commas.

This vagueness must be tolerated, as it is unavoidable, because 'socialism is a hat that has lost its shape because too many people have worn it'. Similarly, the term 'Green' (also 'ecological' or 'environmental') has become too vague, because too many people have, as the Germans say, put on a Green coat. There are today green Greens, red Greens, pink Greens, deep Greens, light Greens, and so on.

The situation described above has given rise to a great confusion and lack of perspective. 'Socialism' has failed; radical environmentalism has failed to achieve anything worth mentioning; but capitalism has also failed and is failing increasingly to solve any of the problems of the world. One need think only of the vehemence with which the ruling élites of even the richest capitalist countries are today trying to dismantle the welfare state. They are not even justifying this with the argument that it has to be done in order to protect the environment.

This confusion must be removed and a clear perspective must emerge if we want to try to save the biosphere from total ruin, and humanity from more hunger, exploitation, oppression, wars, civil wars, and ethnic cleansings. Creation of a tolerably good human society and an ecological economy are no longer matters for dreaming: they have become necessities for survival. Through this book, I wish to contribute to this vital process. I have tried to remove confusions and create clarity. And I have tried to present a perspective.

Certainly, there are many people who do not think that they are suffering from confusion or lack of a perspective (I am one of them). But I think many of them are suffering from some illusions. One very popular illusion is that further development of science and technology and more intensive use of both would enable humanity to overcome the ecological crisis, save industrial society and enable the South to accomplish sustainable development. Indeed, there is a lot of talk of a technological and/or efficiency revolution. Another popular illusion is that some economic innovations like pollution certificates, ecological tax reform, and so on, would make the transformation of present-day capitalism into ecological capitalism possible. I have explained in chapters 4 and 5 why I consider these to be illusions.

In both of these illusions, the existence of a basic contradiction between ecology on the one hand, and industrial economy and capitalism on the other, is indirectly admitted. Otherwise one would not need a 'revolution' to reconcile them and overcome the ecological crisis. Leftists of all sorts add to this the basic contradiction between the spirit and logic of capitalism on the one side and justice and social welfare on the other. The solutions most people offer today are commonly expressed in the concepts 'eco-social market economy' and 'ecological restructuring of industrial society'.

Because I consider these solutions to be illusory, I have proposed a different solution in chapter 6. The solution I have offered can be called 'radical eco-socialism'. It will be neither a market economy, which is just a euphemism for capitalism, nor an industrial society.

There can be three radical-ecological objections to the term 'eco-socialism'. Firstly, the term was in the 1980s deployed by social democrats and other leftists, who opportunistically merely added some ecological demands or analytical elements to their otherwise traditional social-democratic or socialist positions, which gave rise to many inner contradictions in their programmes, or they sincerely but naively believed that an ecological industrial society was possible. Secondly, the theories of all varieties of socialism demand rapid development of productive forces and a high degree of affluence, considered to be essential elements of policy of a (transitional) socialist or social-democratic regime and indispensable foundations of a future socialist society. Both elements are opposed to the requirements of a really ecological economy. Thirdly, it can be argued that the task today is not one of fighting against capitalism for a socialist society, but of ensuring the survival of the human species or of the whole biosphere. This is the sense in which some early radical ecologists coined the slogan: 'We are neither on the left nor on the right, but in front'.

I share the first two objections, but reject the third. Of course, one may say: what is in a name? That is right. But my conception retains so many elements of the socialist ideal that it would be tantamount to deception if I were not to call it 'eco-socialism'. Moreover, I am convinced that unless we can solve the social question in a socialist (not welfare state) manner, we have no hope of overcoming the ecological crisis.

Some radical ecologists have argued, together with protagonists of an eco-social market economy, that the discussion must be de-ideologised. But I think that is neither possible nor desirable. I am convinced that capitalism has a compulsive orientation towards growth inherent in its logic. There is in its logic no place for justice, equality, fraternity, solidarity, compassion, morality, or ethics. That is why the choice between capitalism and socialism has not become irrelevant; it must be faced.

In fact, I believe it has become more relevant with the emergence of the ecological crisis. Indeed, it is my thesis – I shall elaborate it in chapters 5 and 6 – that a truly ecological economy can function only in a socialist socio-political set-up, and a truly socialist society cannot but be truly ecological. While there is an ineluctable and unresolvable contradiction between capitalism and industrial economy on the one hand and the requirements of a truly ecological economy on the other, there is no contradiction between socialism and a truly ecological economy if the former can be conceived of as a non-industrial society, which I believe is possible. It is also my thesis that some variants of socialist methods of organising the economy – some forms of public ownership of the natural resources and means of production, planning, rationing, and so on – would be essential for a peaceful and orderly transition to an ecological economy.

But the above is not the only reason why I am an advocate of ecological socialism. Milton Friedman, the famous pro-capitalist economist, said: 'If the free market economy were not the most efficient system, I would even then want it – because of the values it represents: freedom of choice, challenge, risk' (quoted in Heuser 1993: 101–102). I reject eco-capitalism, not only because it cannot function, but also and mainly because of the values capitalism represents: exploitation, brutal competition, worship of mammon, profit and greed as motive. And I am for socialism mainly because of the values it represents: equality, co-operation, solidarity. Freedom and democracy are compatible with these values, although they did not exist in the 'socialist' regimes we have experienced up to now, but they are not compatible with the values of capitalism, especially not with inequality in wealth and power. 'Socialism' is dead, but not socialism. Socialism has a future. It must, however, first learn the ecological lesson.

It should now be clear why the ecology movement in the Western capitalist countries failed. In short, it bumped up against the inherent growth-logic of capitalism and industrial economy. But why did 'socialism' fail? I am not satisfied with the answers I have heard so far. It is my thesis that 'socialism' failed in the USSR and Eastern Europe mainly because it encountered the limits to growth. I have read many descriptions of the environmental destruction in the 'socialist' countries, and also some references to their resource problems. But everywhere the causal connection between the limits to growth and the failure of 'socialism' is missing (see, for example, Magdoff & Sweezy 1990). In chapter 2, I present material that proves my thesis.

There was, in my opinion, a second major cause of the failure of 'socialism'. It may be generally called the human factor or the subjective factor (limits to growth being an objective factor). The communist parties of the 'socialist' countries failed to create the 'new man', whose behaviour was to be motivated and guided by a new, socialist/communist ethics, which, they knew, was essential for the success of their project. The project was not only to develop the productive forces and to create an affluent society. The communists also wanted to create a classless society, a morally superior society. They failed because it was and is a very big and very difficult project. I address this point in chapter 3.

The 'socialist' regimes of the USSR and Eastern Europe could have continued (not succeeded) in spite of the economic stagnation if the moral fibre of their communist parties and, with it, that of their societies had remained socialistic. And they could have continued in spite of moral degeneration if economic growth and a rise in the standard of living had continued at a satisfactory rate. But towards the end, neither condition was fulfilled.

Why the two conditions could not be fulfilled can also be generally explained in terms of what may be called economism, which was an integral part of traditional Marxist socialist theory. One element of economism is the belief that continuous economic growth is possible and necessary. The other element is that affluence is necessary for a good life. Because of this economistic thinking, 'socialists' never considered their economic growth rate and the achieved degree of affluence to be satisfactory. Moreover, by the early 1980s, for reasons elaborated in chapter 2, economic stagnation set in. And since affluence could not be achieved for all citizens, the ruling élites started granting themselves privileges which could not be justified, and they indulged in corrupt practices in order to be able to afford the good life that their ideology promised. They became morally degenerate. I have dealt with this matter in some detail in chapter 3.

Apropos of economism, there is (was) another aspect of it, in the West as well as in the USSR and Eastern Europe. Both Western Sovietologists and 'socialist' reformers understood the failure of 'socialism' as merely the failure of its economy. Moreover, they saw the explanation of this failure only in the 'functional deficits' of and the 'functional blockades' in the economic system in the narrow sense, that is, centralised planning, command–administrative methods of management and control, and state ownership of the means of production, all of which allegedly killed initiative and motivation.

As already stated, I believe, in contrast to them, that the major causes of the failure of 'socialism' were the limits to growth and moral degeneration. The failure of a social system can have several aspects. And it can be partial. A social system can be economically (in the short or medium term) successful, but morally degenerate. The USA, for example, is economically apparently very successful. But US society is full of crime, violence, corruption, psychological suffering, drug addiction, unemployment, homelessness, and so on and on. South Korea achieved rapid economic growth, but at the price of dictatorship, intensive exploitation of the working people, and widespread corruption. Even in the rich capitalist countries, capitalism has failed in terms of the quality of the society.

I do not deny the role played by the functional deficits and blockades referred to above in the failure of 'socialism'. They did play a role, but a minor one. Even if these deficits and blockades could have been removed, 'socialism' would have failed – both in the narrow economic sense, because of the limits to growth, and in the broader moral sense, because of the moral degeneration of the élite as well as of the people. Nevertheless, I have devoted some space in chapter 3 to these functional deficits and blockades because I think readers should know the difference between my analysis and that of standard critics of 'socialism'.

Although I have written about the causes of failure of 'socialism' in general, I have analysed chiefly the experience of the USSR. This is not scientifically correct procedure, but more was impossible owing to lack of time. We find ourselves today at a crucial historical juncture. Many developments, including intellectual developments, of long-term global significance are being initiated today – either consciously or by default. The things I have to say must, therefore, be said today and not in five years' time. For a political essay like this, it is enough to analyse the experience of the USSR. The other 'socialist' countries of Eastern Europe had, after all, copied the Soviet system, and their experiences, at least in the economic sphere, were very similar. The experiences of China and of Cuba are very different and should be analysed separately.

Analysing the experience of the USSR is important, not only for finding the causes of failure of 'socialism', but also for visualising a future socialist society. For many of the difficult problems, issues, and choices that will have to be faced in a future socialist society were revealed in the Soviet experience. They were even discussed intensively. The early history of the USSR is particularly important in this regard. The lessons that can be drawn from there are very valuable – even though a future socialist society will have to be built under conditions of limits to growth, of which the early Soviet revolutionaries had not an inkling. Chapters 2 and 3 are devoted to this analysis; chapters 4, 5, 6, and 7 are devoted to the issues of today and of the future.

It is not as if the problems that humanity faces today and will face in the foreseeable future relate only to the two questions: (a) how to overcome the ecological crisis, and (b) capitalism or socialism. There are many other problems and issues in this world. They can be broadly classified under two headings: (1) those concerning the relations of humans to the rest of nature, and (2) those concerning the relations of humans to each other. Under (1) come questions of animal rights, vegetarianism or meat-eating, extinction of species, wilderness, natural resources, size of the human population, deep ecology versus anthropocentrism, stewardship or exploitation of nature for the benefit of humans, the extent of humans' right to utilise the rest of nature. And perhaps more. Under (2) come questions of man–woman relationships, the relations between generations, education, parent–child relationships, the relationship between the various peoples and cultures of the world, racism, minority rights, emigration, war and peace, the state, political organisation, civil society, ownership of means of production, economic organisation, distribution of the national income, equality and justice. And more.

Directly or indirectly, all these problems and issues are relevant to the broad twin goals of overcoming the ecological crisis and creating a tolerably good human society, which is impossible unless and until the ecological crisis is overcome. As young socialists we learnt that everything is connected with everything else. In the ecology movement we have learnt the expressions 'holistic thinking', 'thinking in connections' ('*vernetzt denken*' in German), 'think globally and act locally'. My experience is, however, that after the end of the international communist movement – which was a connected, holistic phenomenon, having an overall theory, analysis, programme and strategy – the movements and efforts to attain the twin goals became disconnected, separated. The activists became specialists, ceasing to bother about the whole. The broad twin goals were forgotten, and attaining partial goals became the main purpose. Not surprisingly, they often worked at cross purposes, and produced contradictory programmes and demands.

To give an example, the Greens of West Germany solidarised with the trade unions in the latter's struggle against the capitalists. So they supported the latter's demand for higher wages and less work. At the same time, they spoke of reducing production and consumption. The trade unionists requested the Greens to support them in the name of social justice, but at the same time they abused the Greens and the ecology movement for being anti-industrial (see Sarkar 1993; 1994). To give another example, many women's rights activists oppose any population control policy coming from anywhere with the argument that only the individual woman has the right to decide how many children she produces. At the same time, they demand that the state or society or community must do everything to make the children happy, and feed, clothe and educate them. In addition, they demand that the environment must be protected, but they refuse to recognise the continuous growth of the human population as a serious problem (for a discussion of the contradiction, see Sarkar 1993a).

We therefore need an overall theory, an overall analysis. Partial theories and analyses will produce only half-truths and half-solutions, which might be even worse than ignorance and no solution. I have the feeling that most books and articles on the relevant issues have dealt with only a part of what is actually a whole. I think a synthesis of the partial truths, insights, and solutions is urgently necessary, essential for the movements. And we need a comprehensive movement.

If we carry out this synthesis, however, we shall see that many of our very popular theoretical positions and concrete demands are actually untenable, insupportable. We must then have the courage to say that we were wrong and to give up our popular but erroneous demands.

I think such a synthesis is possible, and I have tried to do it here. It is difficult to achieve in the limited space of one book, but perhaps I have succeeded in producing at least the framework for it. I hope this book contains a useful discussion of the major problems and issues. It may then be a good introduction to the subject. It must, however, be added here that I have attempted only a theoretical synthesis. A practical, concrete synthesis – that of all the relevant, hitherto separate movements – is the task of activists in the field.

In chapters 5, 6, and 7, on the basis of my analysis, I have presented my conclusions as regards the essential features of an ideal human society of the future. One might object that it is not yet time to come to conclusions. But we must act today, otherwise we shall ruin everything and cause immense suffering. But to act, we need conclusions, even if they are provisional. We shall never know everything. Many questions will remain unanswered for ever. All our actions must necessarily be undertaken on the basis of incomplete and uncertain knowledge.

SOME BASIC STANDPOINTS

To produce the framework of a synthesis, an overall analysis, it is necessary first to go into some fundamental issues and state one's basic standpoints. Only then can one take up more concrete and practical issues. If there are differences, uncertainties, or lack of clarity on these fundamental issues, then a discussion may be very difficult. In the following, I have presented my basic standpoints on such issues.

Gaia, deep ecology and anthropocentrism

I once heard a very pessimistic ecologist say: 'In the history of the Earth, so many species have disappeared. What does it matter if the human species also disappears?' Nobody was shocked at this rhetorical question. It seemed as if everybody agreed that it was a very understandable point of view. But it leads to a further question: if it does not matter, why are we active in the ecology movement or the peace movement or any other movement for a better world?

Logically, one possible implication of the Gaia hypothesis of James Lovelock[2] is the relative or even total unimportance of the human species for the survival of the biosphere. It is not as if we humans have to save it. Lovelock writes:

> It may be that the white-hot rash of our technology will in the end prove destructive and painful *for our own species*, but the evidence for accepting that industrial activities either at their present level or in the immediate future may endanger the life of Gaia as a whole, is very weak indeed. (Lovelock 1987: 107–8)

Lovelock believes that the automatic control system of Gaia has enabled her to survive many ecological catastrophes in the history of the Earth. So Gaia would also withstand all the pollutions that we humans inflict on her. Gaia will simply make the necessary changes in the biosphere. The only question we cannot answer today is whether the human species could also have a place in that changed biosphere.

The point that emerges from the above is that if we are active in the ecology movement, then there is no need to think that we are active in order to save *life* on Earth. We want to save humans and the human species and the present member-species of the biosphere. When we talk of protecting the environment, we mean protecting the present environment in which we humans are at home, feeling more or less well. This may also be, who knows, the only environment in which the human species can exist. Gaia would regulate everything and maintain her own homoeostasis with or without us. But that is a very long-term process. In the short term, millions and millions of humans may suffer. Since we are humans, our main (not our only) concern is the suffering of humans – all the more so if Gaia is capable of looking after herself. This might be called anthropocentrism. But the reason for this anthropocentrism is very simple. It is because we are humans that we are anthropocentric. No philosophising is necessary, and no religion or mythology – no species wants its own extinction or increased suffering.

So the Gaia hypothesis is very interesting for science, and some predictions and tests made on this basis have confirmed it (Lovelock 1990: 102). But in matters relating to us humans and our societies, it cannot replace anthropocentrism.

In the ecology movement, however, there are two related streams of thought and action that strongly criticise anthropocentrism: the animal rights movement and the deep ecology movement.[3] The main concern of both of these is the protection of the species between Gaia and the humans, species that are victims of human dominance over nature.

Theoretically, we have no reason at all to suppose that we humans are in any way superior to other species. We simply have different characteristics. Andrew McLaughlin has summarised Paul Taylor's argument as follows:

> [A]rguments for anthropocentrism ... beg the question at hand, assuming some human characteristic as superior in order to prove that superiority. For example, he considers the Greek idea that man is a rational animal and shows how the idea itself is used as evidence that 'therefore' humans are superior to other creatures. But this criterion is, at best, a human consideration about the best life for humans. It provides no justification for applying such a criterion across all forms of life. The human capacity for reason is no more a justification for a value hierarchy among all life than is the cheetah's speed or the eagle's vision. (McLaughlin 1993: 155–6)

Lovelock argues: 'The Gaia hypothesis implies that the stable state of our planet includes man as a part of, or partner in, a very *democratic* entity'. So, 'from a Gaian viewpoint, all attempts to rationalize a subjugated biosphere with man in charge are as doomed to failure as the similar concept of benevolent colonialism' (Lovelock 1987: 145; emphasis added).

However, two particular characteristics of humans are of great significance. Firstly, humans are one of those very rare species in which an individual kills members of its own species. A tiger fights another tiger over control of a territory. Neither, however, wants to kill the other. The weaker tiger withdraws and that is the end of the fight. But it is very common that humans kill humans – not for eating, but for many different reasons. Secondly, in spite of this abominable characteristic, humans are also the only species that acts against natural selection. We generally look after the old and the sick, we try to make peace after every war, we often regret the killings, we allow the weak among us to produce children, we feed the destitute, and so on. Of late, we also want to protect endangered species. It is only the extent to which we do these things that is, for many of us, not satisfactory. Other animals generally do not do these things. There are cases of mutual aid within other animal species, but only the human species tries to protect other species.

The first of these two special characteristics might indicate that in at least one important respect we are less developed, less 'civilised' than most other species. I think, practically speaking, we would have achieved a lot if we could create a world in which humans at least do not kill humans. Simultaneously, we could also try to create a world in which no human exploits or oppresses another human, no people exploits or oppresses another people. These goals are limited, they relate only to humans. We are appealing to humans to respect what may be called the 'species limit'.

But the second characteristic signifies that we are already capable of rising above self-interest and above our species interest. It is not nature that demands of us that we should protect other species, but our own sense of moral duty. If we think that we humans are only one among the many species in nature, equal to the others, then we do not have any extra moral duty to protect other species, a duty which does not exist in the rest of nature. This shows that we humans are not just another species, but quite a different one, although not superior in value to the others.

Actually, one can point out an inner contradiction in deep ecology. Its protagonists consider other species to be equal to humans in value. The first of the eight basic principles of the Deep Ecology Platform reads:

> The well-being and flourishing of human and nonhuman Life on Earth have value in themselves (synonymous: intrinsic value, inherent value). These values are independent of the usefulness of the nonhuman world for human purposes. (quoted in McLaughlin 1993: 173)

But while they cannot think of killing humans, they can and do allow the killing of animals. Arne Naess has noticed the problem. He writes:

> When we attempt to live out our relationship with other living beings, in accordance with such a *principle of equal rights* of all fellow beings, difficult questions naturally arise. ... Our apprehension of the actual conditions under which we live our own lives ... make it crystal clear that we have to injure and kill, in other words, actively hinder the self-unfolding of other living beings. (Naess 1990: 167; emphasis in original)

It is not as if we kill animals only because of hunger. We eat meat despite the fact that we could survive as vegetarians. We also enslave animals in order to use their labour power for our comfort.

The contradiction can easily be resolved if we say that we want to protect every species but not every living being. I suppose that is the position of most deep ecologists. But in that case they are indeed making a value-differentiation between humans and other animals. They do not mind killing animals, but they cannot think of killing humans. Similarly, if a house is burning and we know that a human and a dog are trapped in it, even a deep ecologist or an animal rights activist would save the human before the dog, if only one of them could be saved. This is not very far from anthropocentrism.

In response to the criticism against anthropocentrism, Christian ideology has adopted the concept of stewardship in place of dominion over other species. Many Marxists and other humanists today also accept this criticism directed against them. This has had positive effects on policy. But the point is that even when our measures to protect the other species are not motivated by our scientific, aesthetic or emotional needs or interests but by a pure sense of moral duty, it is we who decide what our moral duty is, it is we who are granting the other species the right to survive.

It seems, therefore, that there is no escape from anthropocentrism. But I think the dispute can be resolved, at least theoretically, if we introduce a differentiation between 'anthropocentric' and 'anthropic'. Any decision or act of ours that goes against our own needs and interests, and favours other species, is a *non-anthropocentric* one. None the less, it is an *anthropic* decision or act.

The question as to whether it is good or bad for Gaia that we act against natural selection cannot be answered. Maybe it is immaterial for her. One thing is clear: only at the beginning of our evolution were we really equals of the other species. At present, this equality is only a wish and a theory. In practice, it is only a moral duty of ours. Even so, let us not arrogantly think that the whole of the rest of nature is at our mercy. The enemy bacteria that we thought we had eradicated – those that cause cholera, malaria, plague – are all coming back, and we have to kill them, otherwise they will kill us.

Of course, the theoretical or philosophical question will remain: should we protect other species because it is in our own interest, or because we respect their rights, recognise their intrinsic value, and think of them as equals? The question need not be answered immediately. But practised anthropocentrism, the extinction of species caused by humans, must be stopped. I guess almost all thinking humans agree today that it is in our own long-term species interest, if not also in our immediate material interest (medicine, agriculture), to protect other species. If we thereby also bring in a purely philosophical–ethical consideration, it only strengthens the movement. The goal is the same. Bill Devall and George Sessions, well-known deep ecologists, have said something similar:

> We suggest that humans have a vital need for wilderness, wild places, to help us become more mature; but beyond our psychological needs, wilderness is the habitat of other beings which have a right to live and blossom for themselves (inherent value). Thus on both the grounds of self-realization and biocentric equality, the wilderness issue as public policy decision and as re-creative experience assumes great importance from a deep ecology perspective. (Devall & Sessions 1985: 111)

One question, however, must be answered immediately: in what kind of social and economic order is the possibility and probability greater that the above realisation of all thinking humans will become an important content of policy – of governments and of economic units? I am convinced that it will not be possible in a capitalist social and economic order. The ecology movement cannot avoid the question any longer.

One Earth and one world

In a certain sense, we humans are not anthropocentric enough in our relations to each other. Not only as a planet, but also ecologically in the Gaian sense, the Earth is one. But the world is not. I generally criticise the positions of the Brundtland Report. But here I would like to quote a passage for which I have nothing but praise:

> The Earth is one but the world is not. We all depend on one biosphere for sustaining our lives. Yet each community, each country, strives for survival and prosperity with little regard for its impact on others. Some consume the Earth's resources at a rate that would leave little for future generations. Others, many more in number, consume far too little and live with the prospect of hunger, squalor, disease, and early death. (WCED 1987: 27)

If we were true anthropocentrists, we would think of ourselves mainly as humans, and the interests of the whole of humanity would be our chief concern. In fact, however, we think and act mostly as Indians or British or Russians, etc., when we are not thinking and acting totally individualistically. We are still so far away from the ideal of one world that mere tribal solidarity is considered to be great.

The conflict between our national/tribal loyalties and the logical practical consequences that must be drawn from the realisation that the Earth is one was demonstrated at the Earth Summit in Rio in 1992, which failed to produce any result except much paper. Even the smaller European Union still has enormous difficulty in functioning as one unit – even in areas of common ecological concern – because of the national loyalties of its members.

I think that development in this area is taking place in the wrong direction. Since the rise of the working-class movement in Europe in the 19th century, the great majority of activists in the left and new social movements have drawn inspiration from the ideals of internationalism and humanism. But of late, even many well-meaning people in those movements are propagating ideas that contradict these ideals. They want to separate people from each other – in the name of difference, diversity, tradition, and cultural or ethnic identity. The idea of one world has become anathema to them. And at the level of practical politics, two opposed trends exist side by side: the older trend of economic globalism (which existed in both capitalism and 'socialism') is continuing and becoming stronger through the growth of world trade and the internationalisation of capital; at the same time, multi-ethnic political entities such as the Soviet Union and Yugoslavia have been broken up violently. Even in a small, linguistically and religiously homogeneous nation like Italy, the rich North wants to separate from the 'poor' South, and this trend is visible in many other countries.

'Holism', 'think globally', 'everything is connected with everything' – these are popular catchwords or slogans. But there is a question: can we be holistic in our thinking and local patriots or single-issue activists in practice? If we are either of the latter, then isn't our holism an empty slogan? I think we have to act locally upon many problems, simply because we live and work at a particular place and cannot be in different places simultaneously. But since everything is connected with everything, policies and programmes, on the basis of which we act locally, must be informed by and based on the principle and ideal of one world – because the Earth is one and we humans are but one species.

One logical consequence of holistic and anthropocentric thinking is internationalism. There cannot, therefore, be two different visions and two different kinds of basic policy, one for the North and the other for the South. Only the details can be different. I am saying this because it is quite common, among leftists and socialists, to think that a socialist society would be based in the North on an advanced industrial economy, and in the South on a small-is-beautiful subsistence economy. If an advanced industrial economy is good for the North, why should it be bad for the South? And if small is beautiful for the South, why should it not be beautiful for the North too? There is no automatic solidarity, as leftists believed earlier, between the working classes of the various countries of the world. This solidarity is, however, essential; it must be created if we want to avoid both ecological and human catastrophes.

Paradigm shift

Thomas S. Kuhn wrote his book, *The Structure of Scientific Revolutions*, in 1962. His key concept was that of the 'paradigm'. Since then, the terms 'paradigm' and 'paradigm shift' have also been found useful in areas less exact than the natural sciences.

A paradigm can operate in large or small contexts. In the area concerned, a paradigm determines and guides entire patterns of thought. For example, the Ptolemaic geocentric hypothesis was the paradigm in astronomy for centuries. Later it was replaced by the Copernican heliocentric hypothesis, which could explain not only the phenomena that the geocentric hypothesis could explain, but also the anomalies which the latter could not explain or eliminate. That was a paradigm shift.

Some paradigm shifts can be revolutionary. The conviction that the Earth and humans stood at the centre of the universe could not but influence religion, philosophy and ethics. It perhaps also influenced the other sciences. When Copernicus showed that the Earth was just a small planet rotating around the sun, this radically different way of looking at reality could not but have revolutionary effects on the whole of human thinking.

The earliest use of the terms in social sciences that I know of is that made by Herman E. Daly, who, in 1973, saw 'an emerging paradigm shift in political economy'. Daly formulated it in two questions: 'Shall we conceive of economic growth as a permanent normal process of a healthy economy, or as a temporary passage from one steady state to another? Shall we take the flow of income or the stock of wealth as the magnitude most directly responsible for the satisfaction of human wants?' (Daly 1973: 2).

The first use of these terms in serious *political* discussions that I have come across is that made by the American Marxist Paul Sweezy in 1979, when he referred to anomalies in the behaviour of the apparently very strong 'socialist' societies. By analogy with the crisis that arose in the geocentric paradigm in astronomy through the accumulation of anomalies observed in the sky, Sweezy spoke of a crisis in Marxian theory, which at that time served, at least for all kinds of leftists, as a paradigm for the science of history and society. Let me quote him extensively:

> ... we are finding more and more anomalies in the Kuhnian sense, i.e. deviations between observed reality and the expectations generated by the theory. ... For Marx, socialism was a transitional society between capitalism

> and communism. While he purposely refrained from drawing up blueprints, there is no doubt about what he considered the most fundamental characteristics of communism: it would be a classless society, a stateless society, and a society of genuine ... equality among nationalities, races, sexes, and individuals. These goals would certainly be very long-term in nature and might never be fully achieved. But just as certainly they establish guidelines and rough measuring rods. Only a society genuinely dedicated to these goals and shaping its practice accordingly can be considered socialist in the Marxian meaning of the term.
>
> Now, ... the generally accepted Marxian interpretation of modern history leads us to expect that capitalism will be overthrown by proletarian revolutions, and that these revolutions will establish socialist societies. The theory, in fact, is so taken for granted as a reliable clue to what is happening in the world that every society which originates in a proletarian (or proletarian-led) revolution is automatically assumed to be, and identified as, a socialist society.
>
> And this is where the anomalies begin. None of these 'socialist' societies behave as Marx – and I think most Marxists up until quite recently – thought they would. They have not eliminated classes except in a purely verbal sense; and except in the period of the Cultural Revolution in China, they have not attempted to follow a course which could have the long-run effect of eliminating classes. The state has not disappeared – no one could expect it to, except in a still distant future – but on the contrary has become more and more the central and dominant institution of society. Each interprets proletarian internationalism to mean support of its own interests and policies as interpreted by itself. They go to war not only in self-defence, but to impose their will on other countries – even ones that are also assumed to be socialist.
>
> All of this, I think, is now fairly obvious, and of course it is raising havoc among socialists and communists. I think it is no exaggeration to say that by now the anomalies have become so massive and egregious that the result has been a deep crisis in Marxian theory. (Sweezy 1979)

Much water has flowed under the bridge since 1979. 'Socialism' has collapsed – not because of war or civil war, but because of its internal problems. It imploded. Sweezy's list of anomalies could not naturally contain this collapse. But what is very strange is that this list did not contain the two most massive and important anomalies: capitalism was/is continuing to develop the productive forces; 'socialism' was fettering them. Both are completely contrary to what was expected on the basis of the Marxian paradigm. Even by the middle of the 1960s, these two particular anomalies had compelled the leaders of the USSR to try out economic reforms.

Moreover, Sweezy only made a statement of facts, he did not offer an explanation for the anomalies he listed; he said only that Marx and Engels could not be held responsible for them. According to him,

'Marxism works as well as ever ... as a way of understanding the development of global capitalism and its crisis'. In this crucially important sphere, the listed anomalies in the behaviour of 'socialist' societies had, according to him, 'no bearing on the validity of Marxism. ... The part of Marxism that needs to be put on a new basis is that which deals with the post-revolutionary societies (with which, of course, Marx and Engels had no experience)'. He thought that this was the only way Marxists could 'eliminate the disturbing anomalies' (ibid.).

So the Marxian paradigm was 'saved'. The anomalies were 'eliminated'. But we needed and still need an explanation of the anomalies, and we need an explanation of the total collapse of 'socialist' societies, with or without Marxian theory.

A paradigm is supposed to provide the basic explanation of a whole range of observed phenomena, not just of a selection of them. Or should we consider Ptolemy's geocentric paradigm to be valid because it can explain some observed movements of the heavenly bodies? It is clear that Marxian theory is no longer a satisfactory paradigm in the Kuhnian sense. It seems to me that a new paradigm is already visible. It simply has to be recognised as one and considered seriously.

The new, emerging paradigm is based on the thesis of limits to growth, so I propose to call it the limits-to-growth paradigm. I think it can provide the basic explanation of more observed phenomena than any other. Marx and Engels were not aware of the limits to growth. Although Malthus had dealt in the 18th century with the population problem in relation to the growth of food production – an important aspect of the limits-to-growth paradigm – Marx and Engels and all their disciples vehemently rejected his theory. The deeper aspects of the present-day crisis of capitalist societies – not the usual recessions and slumps in business cycles but those aspects which have something to do with the resource problem, with nature's ability to absorb man-made environmental disruptions, and with the carrying capacity of the Earth (or of a particular country) – cannot be explained within the Marxian paradigm. For that, the knowledge of limits to growth is absolutely necessary.

The Marxist theory of capitalism's crises dealt only with surface-level phenomena: the tendency of the rate of profit to fall, underconsumption, growing impoverishment of the proletariat, and so on. But even at this surface level, anomalies were soon visible. The proletariat did not become ever poorer, and capitalists and other rich people have always been eager, not compelled, to invest their money in trade and industry despite the (alleged) tendency of the rate of profit to fall. So additional explanations had to be brought in to explain away the anomalies and save the paradigm: counteracting forces, relative impoverishment, for example. But the paradigm could not be saved, for the most important expectation based on this paradigm, namely that developed capitalism would collapse under the weight of its inner contradictions, has not yet materialised. This anomaly can be very easily explained if we shift to the new paradigm. Capitalism has not yet collapsed in the developed countries because the limits to growth have not yet been reached there.

The limits-to-growth paradigm offers a basic explanation of deeper level crises of all present-day societies. That is why it seems to me that this paradigm can also provide a more convincing explanation of the collapse of the 'socialist' societies in the USSR and Eastern Europe, as well as of the failure of the great majority of the countries of the South to catch up with the North in respect of economic development, irrespective of whether they took the capitalist or the 'socialist' path.

A paradigm provides not only the basic explanation(s) of observed phenomena, but also determines, in case of the sciences, the direction of research. In societal matters, it determines the overall direction of policy, and, in the case of practical problems, the direction of actions. In the case of the economy, the earlier and still dominating paradigm, which may be called the growth paradigm, gave the hitherto existing basic direction to economic policies. It is also obvious that the limits-to-growth paradigm would necessitate a radical change, perhaps a U-turn, in all areas of economic policy. And since the economy is the base for all other things in human society, there is no doubt that the paradigm that determines the direction of economic activities has impact on and implications for politics and culture in its broadest social-anthropological sense.

In fact, in societal matters, the two paradigms have corresponding futurological visions. In the passage quoted above, Sweezy mentioned a few expected fundamental characteristics of a Marxian communist society. It is worth mentioning that Marx and most Marxists believe(d) that the real wealth of mankind is free time, that the realm of freedom would come only after overcoming the realm of necessity, and that all of this would be possible only after the productive forces have developed so much that society would function according to the principle: from each according to his ability, to each according to his need. Herman Kahn (1976) and many others – thinking within the same growth paradigm but without communistic ideals – produced their vision of progress towards a high-tech superindustrial society. In contrast to Marx and Kahn, those who have accepted the limits-to-growth paradigm are more concerned with survival, with saving the natural basis of life, especially of human life. And then there are many who try to balance the two paradigms. They visualise sustainable growth or sustainable development and an ecological industrial society.

As far as I can see, the greater part of the ecology movement and of the people sensitised by the ecological crisis at present are advocates of and believers in sustainable development or ecological modernisation of industrial society. That is also the case with those who are still socialists, with the difference that they insist that this must and can be done with socialist relations of production. It seems to me that people with such views, although they think they are performing a balancing act, are still thinking (and acting, if they are in power) within the growth paradigm. Those who have completed the paradigm shift in their thought are very few in number.

I do not mean to say that since the emergence of the limits-to-growth paradigm everything that has been written by orthodox (political) economists has become redundant, or that the adherents of this new paradigm cannot use any of the findings of the others. A paradigm shift never has that effect. We can still operate well with the statement 'the sun rises in the east', although we know that it is not the sun that rises, and engineers can still operate well with Newtonian mechanics, although they have learnt Einstein's theory of relativity. Similarly, in the sphere of the economy, even after we have learnt about the validity of the entropy law in the economic process, we must continue to use, for example, the law of diminishing utility, the law of diminishing returns and its concomitant, the law of increasing costs. But there is no doubt that the futurological visions of the growth paradigm have become obsolete.

Finally, I do not mean to say that limits to growth are the explanation of each and every problem or crisis in the economic and social sphere. A meteorite burns out when it comes too close to the Earth. The explanation of its burning out does not lie in the Copernican helio-centric hypothesis. But why it entered the Earth's atmosphere, why it had such an orbit, can be explained only by this hypothesis. Similarly, it is my thesis that limits to growth were the deeper cause of the failure of 'socialist' societies in the USSR and Eastern Europe. But it does not (fully) explain why, for example, the managers of the state-owned enterprises had to cheat the central planning authorities, or why they had to keep the massive environmental destructions in their countries secret. Nevertheless, the answers to these questions are best sought within the limits-to-growth paradigm.

PESSIMISM OR OPTIMISM?

However much we may seek, we shall never get satisfactory answers to some questions: who we are, where we come from, where we go after death, where we get our consciousness and intelligence from, the meaning of life, etcetera. We have to live without answers to these questions. The only thing we can and should try to do is the next best thing: make our life, our societies, and the world as good as we can. There are some evils and sufferings in the world that cannot be removed. But many can be.

The questions that remain and must and can be answered are: (1) should we accord priority to our long-term species interests or to our short-term (economic) interests if the two conflict (in many cases they do conflict)? (2) In general, how – on the basis of which criteria – should we come to a decision when two interests or goals conflict? (3) What is possible and what is not? (4) If we decide to protect our environment, how do we distribute the sacrifices, which will surely be necessary, and the benefits, which might perhaps also accrue, among individuals and peoples? (5) If we accord priority to our long-term species interests, how do we overcome our egoism, tribalism and nationalism?

These are all difficult questions: even the first question is not yet decided in favour of our long-term species-interests. In a fit of pessimism, Nicholas Georgescu-Roegen wrote:

> Will mankind listen to any program that implies a constriction of its addiction to exosomatic comfort? Perhaps the destiny of man is to have a short, but fiery, exciting and extravagant life rather than a long, uneventful and vegetative existence. Let other species – the amoebas, for example – which have no spiritual ambitions inherit an earth still bathed in plenty of sunshine. (Georgescu-Roegen 1976: 35)

A big difficulty in this regard is the short and narrow horizon of most people in the world. As the authors of *Limits to Growth* wrote in 1972:

> The majority of the world's people are concerned with matters that affect only family or friends over a short period of time. Others look farther ahead in time or over a larger area – a city or a nation. Only a very few people have a global perspective that extends far into the future. (Meadows *et al.* 1972: 19)

Perhaps humans are just another species of animals, nothing more, nothing better. Perhaps one day scientists will find conclusive evidence for that. They may then say that the saints – Gandhi, Jesus, Buddha,

Tolstoy, Thoreau, and others – were only so many exceptions, or mutations which had no luck in the process of natural selection.

But we do not know yet for sure. Maybe, despite much evidence to the contrary, the human species does have, at least, the potential to form a tolerably good society. The saints are evidence of that potential. That is why we must insist: *no giving up yet.*

NOTES

1. The terms *socialism* and *communism* always generate some confusion. From the 1870s to the end of the First World War, the two terms were used in the socialist movement synonymously (cf. Conert 1990: 253). The system, the failure of which we are discussing, was called 'socialism' or 'actually existing socialism' by people and groups who called themselves 'communists' or 'communist parties'. I shall not use the term 'communism' to denote the system under discussion unless it is unavoidable. Marx's understanding of communism (contained in his *Critique of the Gotha Programme*) has, in any case, become an unrealisable vision.

2. Lovelock explains the Gaia hypothesis as follows: 'This postulates that the physical and chemical condition of the surface of the Earth, of the atmosphere, and of the oceans has been and is actively made fit and comfortable by the presence of life itself. This is in contrast to the conventional wisdom which held that life adapted to the planetary conditions as it and they evolved their separate ways' (1987: 152). This 'life itself' has been likened to a self-regulating living super-organism and given the name 'Gaia', the ancient Greek Earth goddess.

3. In short, deep ecology says that natural diversity has its own intrinsic value. That is, it should be maintained and the non-human species – both animals and plants – should be protected, not because they are useful to humans but because of their intrinsic value. To this end, human population must be reduced (see Schwarz 1987: 126–7).

CHAPTER 2

Why the Soviet Model of 'Socialism' Failed (1): Limits to growth and ecological degradation

The realm of freedom really begins only where labour determined by necessity ... ends (Marx 1981: 958–9)

WHAT DOES 'SOCIALISM'S' FAILURE MEAN?

It is not as if 'socialism' failed in 1989. It only collapsed in that year; its failure was evident much earlier. A system can fail only if it has an ideal or at least a high goal, and the degree of its failure is proportionate to the height of the goal. Ten to twenty per cent unemployment, as in Western Europe today, or abject poverty plus dictatorship, as in many countries of the Third World, is almost part of capitalist normality. But lack of freedom and democracy, and a standard of living lower than that of England or France are considered to be sufficient proof of failure not only of 'socialism' but also of socialism.

The Stalinist terror regimes[1] – in which not even the whole Communist Party really ruled, but only a small group of leaders (in some cases, only one person) with the help of the secret police; in which not even communists had a right to say what they thought, let alone workers and peasants – such regimes did not deserve to be called socialist. There are many other reasons that many socialists and communists refused to recognise these regimes as socialist. But the main and concrete cause of the failure of the 'socialist' regimes in the USSR and Eastern Europe was their economic failure. They failed to deliver the goods (literally), they failed to catch up with the capitalist West economically, they could not satisfy the growing desires of their citizens as consumers. Some of them even became highly indebted to Western capitalist regimes: the German Democratic Republic (GDR) was in 1989 on the verge of insolvency in its foreign trade.

Lack of freedom and democracy was the most important cause of dissidence for a minority of citizens, who would have protested and demonstrated even if the standard of living had been much higher. But if the standard of living had been equal to that in the West, these regimes could, without fear, have allowed their citizens some freedoms and democratic rights, if not a multi-party parliamentary democratic system.[2] In the case of the GDR, the regime would not have feared that a large number of its skilled workers would not return after a visit to West Germany. Their economic failure and complete helplessness in finding a solution to their economic crisis was also the reason for their lack of determination to defend their system and later to prevent the political breakdown.

One thing must, however, be stated clearly. The economic failure of these 'socialist' regimes was a *relative* failure, only in comparison to the 'success' of *some* Western capitalist countries. A Third World person like me, who could observe the standard of living in the GDR, was not at all convinced that its economy was in a catastrophic state. A GDR citizen said in 1989 on TV: 'We have everything necessary, we lack only the good things', and a West German TV reporter said that the people in Czechoslovakia lived in relative prosperity. Even the foreign indebtedness of these 'socialist' countries was low compared to that of most Third World countries, and the threat of insolvency could have been removed if they had given up their ambition of catching up with the West. In fact, many Third World countries have in recent times experienced near insolvency, and their regimes did not fall. Similarly, the 'socialist' regimes could also have survived with crisis and stagnation if their goal had not been so high. When, therefore, some people in the West spoke before 1989 of the total failure of 'socialism' as an economic system, it was nothing but psychological warfare.

The dislocation and subsequent collapse – economic and political – and with it the catastrophic fall in the standard of living and the breakdown of law and order in the years after 1989, were not the results of 'socialism', but of ill-advised, unprepared, sudden, chaotic efforts to change the system from 'socialism' to capitalism with political plurality. In the USSR, the breakdown started after 1987, when Gorbachev and his friends tried to impose their *perestroika* and *glasnost* too rapidly. In the other 'socialist' countries of Europe, this occurred after it became clear that the USSR was moving rapidly towards capitalism, particularly with the dissolution of the CMEA (Council for Mutual Economic Aid, formerly COMECON) and its internal, planned economic exchange relations. In the GDR, the economic collapse came after the total opening of the country towards West Germany and the introduction of the Deutschmark as the legal currency. In the USSR, the collapse became total with the dissolution of the federal union.

Before 1989, the average standard of living in, say, the GDR, Czechoslovakia, the USSR and Bulgaria was comparable to, if not higher than, that in, say, Spain and Portugal – two OECD countries. But the USSR was always compared with the USA, and the GDR with the Federal Republic of Germany (FRG). And the 'socialists' themselves wanted it so. Their own yardstick for measuring progress was the standard of living in the *rich* capitalist countries.

However, there is no question that politically, in the matter of gaining legitimation, 'socialism' failed absolutely. The 'socialist' regimes failed to get the stamp of approval of the majority of their peoples. They failed to satisfy an important human need, namely political freedom. In this respect, even some poor Third World countries, such as India and Sri Lanka, were superior to them. A certain degree of political freedom and more or less free elections characterised even these poor capitalist countries.

But the greatest failure of 'socialism', a failure with deep and most serious historical consequences, lay in the area of ideology. Seventy-three years after the October Revolution in the USSR and 40 years after the installation of 'socialist' regimes in Eastern Europe, the people there demanded bourgeois freedoms and more prosperity, and expected them from capitalism, instead of demanding liberty, equality and fraternity and expecting them from socialism.

Although their economic failure was only relative, there is no doubt that by the 1980s they had a real economic crisis, in the sense that even the prospect of catching up with the USA or the FRG in the near future was vanishing rapidly. Academician Abel Aganbegyan, Gorbachev's chief economic adviser in the early *perestroika* period, wrote in 1988, 70 years after the October Revolution, that in terms of productivity the USSR was two-and-a-half to three times behind the USA and two to two-and-a-half times behind other developed Western countries. In Soviet agriculture, productivity was five times inferior to that in the USA. Moreover, since productivity (and – as he forgot to mention – the GNP) was growing in the Western developed countries at an annual average rate of 2.5–3 per cent per annum, Aganbegyan saw no chance of quickly catching up with the USA, despite the planned accelerated increase in productivity in the USSR by an average of 6 per cent per annum to the year 2000. To any impartial observer of the situation in the USSR at that time, this rate must have appeared to be absurdly over-optimistic. But the Soviets were euphoric at the time. Nevertheless, simple arithmetic led Aganbegyan to conclude:

> A certain period of time beyond the year 2000 will be required therefore to

> draw up to, and then possibly to overtake, the USA and other countries in productivity, efficiency and quality in the various aspects of the standard of living of the population. In any case we look with confidence to the future and hope that the 100th Anniversary of the Great October Socialist Revolution in 2017 can be celebrated with the fulfilment of the prophecies of Lenin of higher levels of productivity under socialism than under capitalism. (Aganbegyan 1988: 39–40)

Statistics from 'socialist' countries were always known to be either exaggerated or slightly understated. Aganbegyan also confirmed this. We must therefore quote statistics and comparisons originating in the capitalist West. In 1985, the economic journal *Wirtschaftswoche* of West Germany made a study which compared productivity and real wages in the FRG and the GDR between 1960 and 1984, and made a forecast for 1990. It showed that the gap between the two was widening instead of closing, as shown in table 2.1.

Table 2.1 Productivity and real wages in the GDR as percentages of those in the FRG

	1960	1970	1984	1990 (forecast)*
productivity	70	55	46	35
real wages	78	58	44	35

* The forecasts for 1990 later proved to be correct (cf. Kurz 1991: 94).

That was all very serious for regimes which had claimed that the 'socialist' economic system was superior to the capitalist one, and that it would enable them to overtake even the USA in standard of living, a prospect which was for long ideologically used as the main compensation for absent freedoms. The crisis was there not just because the economies of the USA, the FRG and others were growing faster than theirs: there were concrete problems which also slowed down the growth of these 'socialist' economies. These problems were already evident in the mid-1960s and necessitated various economic reforms after that. But, in spite of all reforms, the problems remained; there was no long-term improvement, and so the radical, large-scale economic and political reforms called *perestroika* and *glasnost* had to be tried in the second half of the 1980s.

LIMITS TO RESOURCES

The reason that all the economic reforms failed to overcome the crisis was that the main cause of it could not and cannot be removed, namely limits to growth.

It was good luck that the world's first 'socialist' revolution took place in the huge USSR, which was and still is full of resources. The enemies of 'socialism' could not strangle the world's first 'socialist' regime economically after they failed to defeat it militarily. And the conception of socialism which sees the development of productive forces and material abundance as indispensable foundations for a socialist society could be tried out in spite of relative economic isolation. No other country could have offered better resource conditions for starting to build a 'socialist' society.

Nevertheless, by the 1960s many Soviet experts began to express the 'fear that the country's richest resources would soon be exhausted. Especially disturbing were indications that much valuable ore and oil were being left in the mines or wells or were being discarded. ... many mines and oil wells in the country had only a 50 percent recovery rate.' Fifty to sixty per cent losses were reported on the extraction of coal, oil, potassium, and natural gas (Goldman 1972: 48).

According to Marshall Goldman, the leaders of the USSR at that time did not seem able to understand what the difficulty was. He explained the phenomenon as follows:

> Apparently what happens is that after a Soviet mine operator has extracted the richest ore, his marginal costs and average variable costs begin to rise. As it takes more units of labour and machinery to extract one unit of ore and oil, the mine director begins to look for another, more easily exploited mine or oil deposit. ... the mine director is tempted to move when his marginal costs at the old site (primarily labour, capital and shipping) begin to exceed the cost figure at the new site. (ibid: 48–9)

It is reasonable to believe that not only the mine operator, but also the planners, who were keen to achieve immediate results, wanted it so.

It is not enough, therefore, to know how many resources, in what quantities, there are on or under the surface. The costs of extracting and transporting them are also a very important factor of economic growth. In a later publication, Goldman even justified the behaviour of the mine operators: 'Given its relative resource endowment, however, the Soviets may not have been all that irrational to squander their abundant raw materials and instead seek to conserve other, more limited factors of production, such as labour or even capital' (Goldman 1983: 60).

In a paper published in 1977, Aganbegyan described the plans and prospects of exploiting Siberia's 'enormous natural wealth' – oil, gas, various kinds of minerals, potentials for hydroelectricity, timber, and so on. He even spoke of 'Siberia's large areas of fertile soil'. Although he also spoke generally of 'that forbidding territory', he was euphoric about the plans and projects. He spoke of 'favourable conditions', and he was convinced

> that all these huge sources of electricity in Siberia are *extremely efficient in economic terms*. The energy produced here is the cheapest in the country. This also applies to all the types of resources referred to above. The energy and raw materials are cheap because of the favourable geological and mining conditions and the high concentration of resources in Siberia. These economic advantages are not lost even when Siberia's energy resources have to be transmitted over a distance of 2–3 thousand kilometres ... (Aganbegyan 1979: 72; emphasis added)

But only eleven years later, in 1988, he wrote:

> Yet more striking are changes in the extraction of fuel and raw materials. In the 1971–75 period the volume of output of the mining industry increased by 25% but only by 8% in 1981–85. This decline in growth was mainly connected with the worsening of the geological and economic conditions of mining. With its large-scale mining industry, ... the Soviet Union is fairly rapidly exhausting the most accessible of its natural resources. To maintain levels of extraction, it is necessary to dig deeper, to discover new deposits and to transfer to less favourable fields. The fuel and raw material base in the inhabited regions of the country is already unable to meet our requirements and in many of them the volume of extraction is declining. It is necessary therefore to discover new deposits in the north and eastern regions, to construct transport links, to create new towns and develop territories and attract the population there. All this, naturally, does not come cheap. As a result the cost of fuel and raw materials is growing.The capital investment allocated is growing especially rapidly and the ecological demands in the development of the mining industry are becoming increasingly strict. (Aganbegyan 1988: 8)

In 1977, the 'geological and mining conditions' appeared to be 'favourable', and the resources the 'cheapest' and 'extremely efficient in economic terms'. In 1988, the 'geological and economic conditions of mining' were 'worsening', and the process of extracting the resources was 'not cheap'.

There were some Soviet scientists who had contradicted the euphoria about Siberia by 1977: for example, Boris Komarov,[3] an ecologist and high scientific functionary of the USSR. Komarov's main objections were based on the environmental degradation[4] wrought in Siberia by mining and other industries. He also explained why, except in its temperate zone, the purely financial costs of economic activities in Siberia are too high. He includes reasons that Aganbegyan did not mention in his 1988 book. On the biological resources of Siberia, Komarov wrote:

> The biological output per hectare of tundra, forest tundra, and northern taiga is dozens of times less than in central zones. This factor largely determines the potential for exploiting resources and the chances for providing the population there a normal life; no technology can increase the biological output to any major extent. (Komarov 1980: 118)

The forest areas of northern Siberia are also unsuitable for large-scale use as pasture for cattle (ibid: 115).

Komarov gave another reason why the opening up of northern Siberia for economic exploitation is too costly. Large houses in permafrost areas often sag or develop cracks. As a result, in northern settlements, hundreds of houses have to be given up every year. He cited the case of a 1,000-km road 'devoured' by permafrost: it sank below mud in the summer and reappeared later here and there, pushed up by frozen water below. Building motorways in the tundra is a problem. After being used for some time, they become impassable mud tracks. Such roads often move to the right or left. In summer, some settlements in the tundra cannot be reached at all, as the roads are by then broken up and marshy. As of 1977, modern construction technology knew no answer to these problems (ibid: 113).

Let us look a bit more closely, by way of example, at the production of oil, today's most important mineral resource. By the mid-1980s, Soviet leaders were talking of a 'tight situation in the fuel balance' of the country, and there were reports of falling production and exports of oil (Heuler 1986: 23ff). That is understandable. In 1992, known exploitable oil reserves of the former USSR plus Eastern Europe were estimated to be 8.1 billion tons (Federal Ministry of Environment 1994: 38), and Soviet oil production in 1987 was known to be 624 million tons (Lemeschew 1988: 338). At that rate of production, known reserves were expected to be exhausted in roughly 13 years. What is worse, it was stated in 1988 that the cost of extracting oil had gone up from 3 roubles per ton in some unspecified earlier period to 12 roubles per ton in 1988 (ibid: 330).

In 1994, in a study of the oil corporation BP, it was stated that Russian oil production had been declining for the previous seven years, and that in the territory of the former USSR, it stood at 60 per cent of the highest figure for 1987. The authors commented that the decline of the energy sector in the former USSR was unique in peacetime (*Frankfurter Rundschau*, 27 January 1994). This dramatic fall in oil production was certainly not all due to rapid exhaustion of the resource. The chaos in the country and the failure to make adequate and timely investments were at least partly responsible for it. But, in turn, the lack of investment funds, and perhaps the chaos, were a result of the economic crisis, which, in turn, was largely the result of the general resource crisis testified to by Aganbegyan.

The situation was bad not only in oil, but also generally in regard to fuels and raw materials.

> While 20–25 years ago one rouble of production[5] in the extraction industry required two roubles of capital investment, by 10 years ago this figure had grown to three to four roubles of capital investment and in the 1981–85 period it exceeded seven roubles. Production costs grew for many types of raw materials as did the labour intensity needed for their extraction. (Aganbegyan 1988: 72)

The share of the predominantly raw material based branches (fuel, metallurgy, timber and building materials) in total industrial investment was 33 per cent in 1975, 36 per cent in 1980, and 40 per cent in 1985. And all this, wrote Aganbegyan, was due to the worsening of geological and economic conditions (ibid: 103).

If production of basic resources, particularly of energy resources, becomes ever costlier in real terms, that can only have an adverse impact on the economy in general. Between the five-year-plan periods 1971–75 and 1981–85, the rate of growth of national income fell from 28 per cent to 16.5 per cent, that of resources from 21 per cent to 9 per cent, that of capital investment from 41 per cent to 17 per cent, and that of agriculture from 13 per cent to 6 per cent (ibid: 10). Further (in general), 'the capital output ratio has sharply fallen with every five-year period, on average by 14%. In other words for every rouble introduced into the capital stock there has been less and less production, and less and less good use has been made of it' (ibid). It is no wonder that for many years the economists of 'socialist' regimes characterised their economies as an 'economy of shortages'.

LIMITS TO GROWTH OF FOOD PRODUCTION

Prior to 1917, Russia was a major grain exporting country. From 1909 to 1913, Russia exported on average 11 million tons a year. But in the post-revolutionary period, Soviet grain exports never exceeded 7.8 million tons, a figure reached in 1962, although huge funds had been invested in machinery, irrigation, chemical fertilisers and pesticides (Goldman 1983: 63).

One purpose of the collectivisation drive of 1929–30 was to transfer to industry the surplus produced in agriculture. This purpose was obviously not fulfilled to the extent expected by the planners. There was large-scale hunger and malnutrition, and widespread resistance. The malnutrition was perhaps partly caused by the compulsion to export grain as a part of the strategy of 'primitive socialist accumulation'. But as regards the second half of this century, there is consensus also among Western observers that the basic physiological requirements of the people were satisfied. Indeed, after the death of Stalin, Soviet leaders made conscious efforts to improve agricultural production and the situation of the peasants.

Socialists all over the world were therefore shocked to hear in 1963, and again in the 1970s, that the USSR had to import wheat from Western capitalist countries. In addition, it imported large quantities of meat and dairy products. Import of meat and meat products increased from 225,000 tons in 1971 to 980,000 tons in 1981 (ibid: 66).

Most Western observers have explained this as the inefficiency of collective farming. But that is not the full explanation. What is generally not mentioned is that the population of the USSR grew steadily from 136.1 million in 1922 (Ferenczi & Löhr 1987: 263) to 283 million in 1988 (Aganbegyan 1988: 17). From the mid-1950s, feeding this growing population was no great problem for Soviet agriculture; the grain harvest also increased – albeit with fluctuations and many severe setbacks – from 81 million tons in 1950 to 210 million tons in 1986 (Goldman 1983: 177), enough to satisfy the basic physiological needs of the people.

It is therefore clear that the USSR did not have to import food because there was hunger in the country. But its brand of communist ideology and its leaders had promised 'goulash communism'. So the people demanded ever more goulash – meat and dairy products. The result was a chronic meat deficit. By 1985, anticipated annual per capita demand for meat and dairy products was 62 kg and 325 kg respectively. These were expected to rise by 1990 to 70 kg and 340 kg respectively (Aganbegyan 1988: 17–18). It was this ever increasing demand for meat and dairy products that could not be met by internal production, especially in years of bad harvest. Grain had to be imported as cattle fodder.

Because retail food prices could not be increased for political reasons – Khrushchev tried it in 1962 and it led to protests and riots (Goldman 1983: 77) – and farms had to be given higher prices and other benefits, Soviet agriculture became a heavily subsidised enterprise. (One can, of course, also say that the consumers were given subsidies in the form of

collective farming + population growth

prices lower than the cost of production.) By 1980, annual subsidies to agriculture totalled the equivalent of about US$50 billion, of which US$35 billion were needed to subsidise wheat and dairy products (ibid). From being the main source of 'primitive socialist accumulation', agriculture became a burden on the rest of the economy.

In order to increase agricultural production, Soviet leaders invested heavily in that sector. There was steady improvement, but at a high cost. For instance, in Belorussia, between 1970 and 1974, agricultural output rose by 29 per cent, but costs rose by 44 per cent. That means 1.15 roubles were spent to get 1 rouble worth of produce (Nove 1982: 152). And on average, in the whole country, material costs per unit of output rose in the period 1966–76 by 77 per cent, while payments for inputs from outside agriculture rose by 110 per cent. This was not only because more inputs were used: the prices of some inputs – such as tractor-power, fertiliser, concentrated feeds – had also been raised (ibid).

These problems of Soviet agriculture were not all man-made. The geographical conditions of the country were (and still are) the main limiting factor. The continental and northerly location of the country cannot be changed. Only about 10 per cent of the landmass of the USSR could be farmed, as opposed to 20 per cent of the USA. Land used for all agricultural purposes (pastures, orchards, etc. included) covers only about 27 per cent of the country, compared with 43 per cent of the USA (Encyclopaedia Americana 1982: 428i).

The country's black earth region is, of course, very fertile. But the aridity of the climate, except in western Ukraine, is a disadvantage. Rainfall in the Volga black earth region is only about a half or even one-third of that in the western region and is sometimes less than twelve inches annually. The snow in these areas is of little use to agriculture, since the thaw is too rapid and the moisture, instead of penetrating the ground, runs off and often washes away the soil, forming ravines (Dobb 1993: 40).

The greater part of the rest of the soil is of relatively poor quality. Also some of the virgin land that was brought under cultivation in the 1950s in Kazakhstan was not good as permanent agricultural land, although most of it continues to be cultivated. Goldman reported that in the mid-1950s, the programme brought dust storms to Kazakhstan because it had destabilised the soil cover. He actually spoke of 'disastrous' environmental consequences, which had perhaps been the reason why large-scale agriculture had been discontinued in the area in the early 1930s (Goldman 1972: 171).

In general, the climate is unfavourable. Soviet agriculture had (has) a shorter growing season. It was (is) also highly liable to frost and drought damage, because rainfall was (is) unreliable and the mean temperature fluctuation too great. All this was (is) a result of the continental position of the Soviet growing regions, lacking the moderating influence of nearby oceans (Encyclopaedia Americana 1982: 428i). For all these reasons, its greater size and per capita land availability were of little avail to the USSR.

Limits to technological solutions

Geographical limits can, of course, be overcome to some extent by means of water management projects such as irrigation and drainage, and through chemical fertilisers. But irrigation is also ultimately dependent on precipitation, a limited, though renewable, resource.

It has been estimated that the territory of the former USSR receives 10 per cent of the world's annual precipitation. This appears to be a very good endowment. However, the greater part of this water is available in the northern and eastern parts of the country, while the greater part of the population lives in the western part. Only 18 per cent of the water flow is available to 80 per cent of the population, and in effect only 14 per cent is available to heavily populated, industrialised regions. Moreover, much of it runs off as floodwater, while many rivers are dried up for much of the year (Goldman 1972: 78–80). So the irrigation potential is also quite limited.

In the USSR, the positive effects of the 'hydrotechnological restructurings' of the south-flowing rivers were not satisfactory. In respect of irrigation, compared to the huge area flooded (over 120,000 sq km), the result was below expectation (Heuler 1986: 23ff). In many irrigated areas, the harvest diminished instead of increasing. About half of the area flooded by dams consisted of land that was already being cultivated or could have been cultivated (Goldman 1972: 258). Some observers even believe that in the end more fertile land was lost through irrigation projects than was made arable – not only through the reservoirs, but also through salination, water-logging, and so on.

Soviet hydro-engineers also drained several hundred thousand hectares of swamp. In this case also, the result was disappointing. At least half of the disappointing results were due to soil structures. For instance, draining often dried up sandy soil, which was blown away by the winds to form sand dunes. In the case of peat soil, all existing plants often died because of dryness, and new useful plants needed sprinkling. Moreover, the improved land experienced dust storms. In other areas, the thin humus layer was removed along with the brush roots, and then ploughing brought sand on to the surface (Komarov 1980: 48–50).

As for chemical fertilisers, the capacity of plants to absorb them usefully is limited. Three Soviet scientists wrote that only about 40 per cent of the fertilisers used so profusely in their country could be absorbed by the plants. The other 60 per cent was wasted, and polluted the waters (Heuler 1986).

The water diverted from the rivers for agricultural operations of course failed to flow elsewhere, which also had adverse economic (and ecological) effects. The Aral Sea now has only half the water it used to have. The Caspian Sea is also shrinking – mainly due to the reduction in the flow of rivers into it (the other causes are geological changes, and changes in temperature and rainfall). In case of the Sea of Azov, which is connected to the Black Sea, the water became more saline when the flow of river water into it was reduced.

Reduced freshwater flow combined with pollution from industries and cities built on the tributary rivers have had deleterious effects on fish-life in these seas. The total catch has fallen drastically. The total catch in the Sea of Azov in the 1970s was only one-ninetieth of the figure for the 1940s (Komarov 1980: 38). Soviet geographers warned that the elimination of the Aral Sea would increase the continentality of the climate of the region, which would mean longer winters and shorter, hotter summers. The area might, they said, become a desert with frequent salt-and-dust storms. Both factors would have adverse effects on agricultural land nearby.

Solutions to the various problems caused by the drying up of the Aral Sea and the shrinking of the Caspian Sea were sought in an old dream: making some of the waters of north-flowing rivers flow southwards. The idea was twofold: to supply water to the shrinking Caspian and Aral Seas and to make more virgin land in the arid areas arable. But these projects could not be carried out, because the huge funds needed for them were not available. If they had been realised, these projects would have had far-reaching consequences: large areas of arable land, meadows and forests, and many villages and towns would have been submerged. Economic benefits in the narrow sense would have been doubtful, as eleven members of the Soviet Academy of Sciences showed in their calculations (Heuler 1986). And they would have had incalculable effects on the climate through, for example, the further cooling down of the Arctic Ocean.

The shortcomings of Soviet water projects were not wholly due to the inefficiencies of a planned, 'socialist' economy. Almost everywhere else in the world, large water projects have belied the high hopes set in them. Fred Pearce summed this up in 1992:

> ... virtually none of these projects have met their production targets or lived up to financial expectations.
>
> The common image of irrigation making barren land fertile is wrong. Most land taken over for irrigation projects was cultivated before the engineers arrived. ...The irrigators usually take the best land in order to secure the best financial returns on their substantial investments. Frequently, ... water diversions for large state irrigation schemes dry up productive land downstream. So statistics showing how much greater crop yields are on irrigated land may be highly misleading as a measure of irrigation's success.
>
> 'It is now widely recognised that irrigation and drainage projects are not producing the benefits expected', concluded staff from the World Bank's agriculture department in 1992 ...' (Pearce 1992: 183)

The Soviets did succeed – partly through irrigation projects – in raising agricultural production. But endless growth is not possible anywhere in the world. In the USSR, they almost reached the limits to growth of food production. They could not overcome the ultimate geographical constraints. Bad harvests were, without doubt, partly due to mismanagement that could be attributed to the 'socialist' system. Bad management can be remedied, but not the limits to growth.

THE REVENGE OF NATURE

Cost–benefit analyses have in the past generally neglected ecological costs, and they partly still do so. The cost–benefit analysis of the irrigation projects using the waters of the Amu Darya and Syr Darya is a case in point. The planners did not think of the possibility of the region around the Aral Sea becoming inhospitable for humans as a result of the drying up of the sea. In parts of the irrigated area, later, swamp conditions were created and fish that used to feed on mosquito larvae were eliminated. As a result, there was a rise in the mosquito population and malaria returned (Goldman 1972: 235). There was (is) also the fear that the people of the region might decide to emigrate in large numbers to more hospitable parts.

Economy versus ecology: the 'socialist' position

'Socialists' in particular should have been aware of the ecological consequences of economic growth, for Marx and Engels had given warnings in the 19th century. Marx wrote:

> In modern agriculture, as in the urban industries, the increased productiveness and quantity of the labour set in motion are bought at the cost of laying waste and consuming by disease labour-power itself. Moreover, all

> progress in capitalistic agriculture is a progress in the art, not only of robbing the labourer, but of robbing the soil; all progress in increasing the fertility of the soil for a given time, is a progress towards ruining the lasting sources of that fertility. The more a country starts its development on the foundation of modern industry, like the United States, for example, the more rapid is this process of destruction. Capitalist production, therefore, develops technology, and the combining together of various processes into a social whole, only by sapping the original sources of all wealth – the soil and the labourer. (Marx 1954: 506–7)

And Engels, using the negative environmental effects of deforestation as an example, wrote:

> Let us not, however, flatter ourselves overmuch on account of our human victories over nature. For each such victory nature takes its revenge on us. Each victory, it is true, in the first place brings about the results we expected, but in the second and third places it has quite different, unforeseen effects which only too often cancel the first. ... at every step we are reminded that we by no means rule over nature like a conqueror over a foreign people, like someone standing outside nature – but that we, with flesh, blood and brain, belong to nature, and exist in its midst. (Marx & Engels 1976, Vol. 3: 74–5)

But Marx and Engels were none the less growth-optimists. So they also saw a way out, a way of avoiding nature's revenge. Engels wrote:

> all our mastery of [nature] consists in the fact that we have the advantage over all other creatures of being able to learn its laws and apply them correctly.
>
> And, in fact, with every day that passes we are acquiring a better understanding of these laws and getting to perceive both the more immediate and the more remote consequences of our interference with the traditional course of nature. In particular, after the mighty advances made by the natural sciences in the present century, we are more than ever in a position to realise and hence to control even the more remote natural consequences of at least our day-to-day production activities. (ibid)

This optimism cannot be flawed, except that Engels forgot to say, firstly, that there are also limits to the possibility of this control – limits that are also inherent in the laws of nature – and, secondly, that this control involves costs. If a society does not or cannot bear these costs, then environmental degradation goes on unabated and the negative effects accumulate. If the society does bear these costs, which may be high, then the net profit or benefit from some economic activities may be negative or too little to justify their original labour and resource costs. Here lies a fundamental contradiction between (industrial) economy and ecology. I shall take this point up again in chapter 4. For the present, let us see what Soviet theorists and other leading commentators said and did with regard to this contradiction.

Writing in 1976, G. Khozin conceded that all industrially developed countries unavoidably have environmental problems. But he criticised the authors of the first two Club of Rome studies for 'assuming that the crisis phenomena are an equal threat to states belonging to different socio-economic systems', for spreading 'the negative effects of the scientific and technological revolution evenly among all states, irrespective of their social system'. He criticised the alarmists for their 'refusal ... to see the cardinally different approaches to solving the problems ... in different socio-economic formations' (Khozin 1976: 203, 204, 199).

However, on the basis of a thorough study of materials published in Soviet books, journals and newspapers, Marshall I. Goldman came to the conclusion in 1972 that 'environmental disruption' in the USSR was 'as extensive and severe' as in the USA, that the USSR was 'abusing the environment in the *same way* and to the *same degree*' (Goldman 1972: 2; emphasis added). Komarov quoted a member of the Soviet Academy of Sciences, who said in a conversation: 'Our situation is not generally better than theirs, that is, in the USA or in Europe. But for them everything ahead is black, while for us there is at least theoretically a ray of hope' (Komarov 1980: 98).

That was, indeed, the *theoretical* position of all socialists in the 1970s. They saw the roots of the problem in the political-economic system. Firstly, capitalism was (wrongly and dogmatically) seen as a system that fettered the further growth and application of science and technology (that is, of productive forces), and, secondly, capitalists were (rightly) seen as having no logical interest in investing money in the protection of the environment. On both points, the opposite was expected, as a matter of course, from a socialist society. The logic was simple. If all the natural resources and all the means of production are owned by the state, then there cannot be any external costs, as there are in a capitalist economy. For any extra profit made by a state-owned enterprise neglecting to protect the environment would mean losses elsewhere, which ultimately the state would have to bear. The polluter – that is, the state – will not be able to pass off a part of its costs on to society at large.[6] Environmental degradation in 'socialist' societies was therefore seen as a temporary phenomenon that could and would be overcome when the productive forces had grown sufficiently.

Indeed, Soviet leaders also wanted to fulfil the expectations placed in a socialist system. In fact, Lenin gave high priority to conservation, and the USSR was the first country in the world to establish a nature reserve. Between 1925 and 1929, nature reserves grew from 4,000 to 15,000 square miles (McLaughlin 1993: 49–53).

There was a plethora of laws for protecting nature, which Goldman called 'perfection itself' (Goldman 1972: 22). He quoted one B.V. Petrovsky, then head of the Ministry of Public Health of the USSR, who said in 1968:

> In the Soviet Union questions of protecting the environment from pollution by industrial wastes occupy the centre of the Party's and government's attention. It is forbidden to put industrial projects into operation if the construction of purification installations has not been completed. ... The Soviet Union has been the first country in the world to set maximum permissible concentrations of harmful substances in the air of populated areas. (ibid: 23)

Both Goldman and Komarov reported that the maximum permissible concentrations (MPC) of harmful substances in air and water were very stringent, in many cases much lower than in other countries (ibid: 24–5 & Komarov 1980: 21). In addition to the laws, protection of environment was in 1977 made a part of the constitution of the USSR – as a duty of both the state and citizens.

But laws, constitutions and intentions are one thing and economic, political and social realities another. After the 1920s, the conservation movement in the USSR died down gradually. By the early 1950s, the total area of the nature reserves went down from its peak of 48,000 sq miles to 5,700 sq miles (McLaughlin 1993: 53) All sorts of economic activity was going on in them. Laws relating to the protection of the environment were being violated with impunity. Where offenders were punished, the punishments (mostly fines) were so mild that they had practically no effect. Goldman quotes two bewildered Soviet citizens, both prominent. Nikolai Popov, at the time an editor of the journal *Soviet Life*, asked in 1966: 'Why in a socialist country, whose constitution explicitly says the public interest may not be ignored with impunity, are industry executives permitted to break the laws protecting nature?' (quoted in Goldman 1972: 23). Academician Gerasimov asked in 1969: 'What is it in our society with its consistent progress in all spheres of life, that interferes with a rapid advance in such an extremely important field as the rational exploitation of nature?' (ibid).

The bewilderment of Popov and Gerasimov is not surprising. In the framework of their ideology and theories, there was no explanation of the fact that in the 'socialist' USSR environmental degradation was going on in the same way and to the same degree as in the capitalist

USA. But anyone who is free from such ideology and theories knows the explanation: the USSR was an industrial country like the USA, and the fundamental contradiction between the economy and ecology exists in any industrial society.

Moreover, in the USSR, the process of industrialisation began quite late. The Soviets wanted to and had to race to catch up militarily and economically with the already highly developed and inimical West. The economy had to produce and accumulate as much surplus as possible, while there was scarcity in every branch of agriculture and industry. There was not even enough hay for cattle. Every bit of scarce investible surplus was needed for rapid economic growth. There was, therefore, not much left for investing in the protection of the environment, and strict adherence to the relevant laws would have slowed down economic development and the growth of military power.

In fact, even in the 1920s, some nature reserves were simply game reserves, where commercially valuable species were to be propagated. In 1929, it was decided at the conservation congress that the reserves had to be justified in economic terms (McLaughlin 1993: 49–51), and when the five-year plans and accelerated industrialisation started, nature reserves obviously lost support. Economy triumphed over ecology at the level of policy-making.

It is not as if the twin policies of socialism in one country and industrialisation at break-neck speed were the result of Stalin's decision only. Development of science and technology, and unbounded faith in unlimited progress and the unlimited abilities of humans when freed from the fetters of capitalism dominated the *Zeitgeist* in the USSR. This *Zeitgeist* led Soviet intellectuals to forget the words of caution written by Marx and Engels. In 1926, the author Sasubrin said at the first Siberian Writers' Congress (1926):

> I wish that the tender, green breast of Siberia be clothed with the cement armour of cities, armed with the stony muzzles of factory chimneys and chained with the railway lines. The Taiga should be reduced to ashes, thinned out, the steppe may be trampled down. Let all that be, it will be unavoidable. Only on cement and iron can the fraternity of all humans, the iron brotherhood of the whole humanity be built!' (quoted in Rosenbladt 1986: 143)

The poet Mayakovsky saw the industrial city Novokusnezk rising deep in Siberia. In his great enthusiasm, he wrote the following verse on the city:

> 'I know a city will exist,
> I know a garden will bloom,
> When such people exist
> In the Soviet land'
>
> (quoted in Komarov 1980: 23–4)

And Maxim Gorki said in 1937 on the newly constructed canals: 'They are big rivers, my lord, and sane. Before, rivers were mad' (ibid: 143). It was a *Zeitgeist* of changing nature, of ruling over nature like a conqueror. Nature took its revenge. And because the rule of the Soviet people over nature was particularly harsh, nature's revenge here was also more severe than in any other country and in any earlier epoch.

Ecological costs of economic growth

It is surprising that Goldman concluded in 1972 that the USSR was abusing its environment only to the same degree as the USA. Komarov saw the matter differently: the extent of pollution was approximately equal when the volume of production in the USSR was only about half that of the USA. On the basis of his knowledge of the contents of unpublished documents, Komarov feared that by 1984 not just the urban–industrial agglomerations but the whole USSR would inhale more polluted air than US cities (Komarov 1980: 30).

In 1988, according to the Worldwatch Institute, the USSR emitted 18.5 million tons of sulphur dioxide and the USA 20.7 million tons. That is, the USSR had almost caught up with the USA in this regard. But in respect of emission per unit of GNP, the USSR (10 grams) far exceeded the USA (4 grams). In respect of nitrogen oxides, the figures for the USSR (4.5 million tons and 2 grams per unit of GNP) were much better than those of the USA (19.8 million tons and 4 grams per unit of GNP) (French 1991: 96). (This was obviously due to the much lower number of cars in the USSR.)

As regards water pollution, Goldman reported that in 1966, 300,000 cubic metres of untreated sewage was being discharged into the Volga every hour (Goldman 1972: 231). It was estimated that 60–75 per cent of industrial sludge in the USSR was not treated at all. And of all the industrial and household sewage that was processed by sewage treatment plants, only 20 per cent received secondary treatment (ibid: 109).

Komarov's report, first published in Germany in 1978, showed that the situation had not improved despite large expenditures on sewage treatment plants. In some cases, it had even worsened in the meantime. For example, the amount of oil in the waters of rivers and seas appeared to have increased, which was no wonder in view of the continuously rising extraction and use of oil and oil products. The midstream oil concentration in the Volga was constantly 25–30 times MPC, and in the Don 51 times MPC. Moreover, new problems came up, for example DDT in the ground water of Uzbekistan in rather high concentrations. There was real concern about the government's ability to supply drinking water to the people (Komarov 1980: 36–7).

As regards the state of the land in the USSR, Goldman quoted the following data from various sources: 50 million hectares (500,000 sq km) were 'affected' by erosion; from 1958 to 1964, 9 per cent of the country's arable land was eroded, and 24 per cent of the land actually under cultivation 'suffered a similar fate' (Goldman 1972: 170).

Six years later, Komarov gave the following figures, which, although not strictly comparable with those of Goldman quoted above, show a certain worsening of the situation: of the total area of 22 million sq km, 14–15 million are habitable. Of the latter area, in 1977, 175,000–220,000 sq km had been disfigured and devalued through mining, 50,000 sq km were dumping grounds for all sorts of industrial and municipal wastes, 120,000 sq km were submerged under the dam reservoirs, 500,000–550,000 sq km were wasteland and swamps, results of deforestation and forest fires, and 630,000 sq km were former agricultural land that had eroded or become saline, sandy, or full of ravines. In sum, 1.45 million sq km, roughly 10 per cent of the habitable area, had been destroyed.

> And this tenth of our 'pie', which we have devoured in little bites, was the most 'succulent and tasty'. The most suitable for use and, in many cases, the most fertile. We have picked the cherry from the cupcake. Hence we are parcelling out the second, third, and fourth shares as fast as possible. Especially since there are so many more of us, and our needs are growing at a very rapid pace. (Komarov 1980: 130–1)

In addition to this kind of destruction, there was also the usual pollution. Komarov reported that in the mid-1970s, scientists found that agricultural land in the USSR was contaminated with 150 types of pesticides, chemical poisons, and trace elements, of which 90 were compounds of heavy metals (ibid: 47).

Komarov devoted a separate chapter to Siberia. This sparsely populated, huge area, which the Bolsheviks took over in 1917 'in almost the same condition in which it was created by God', was, according to him, 'a unique chance for the Soviet system, a land where new relationships with nature could be built without repeating the gross errors of the past. … Hence the experiment in exploiting Siberia was a pure experiment, a kind of test for the socialist mode of development' (ibid: 112). But the Bolsheviks failed the test. They ruined Siberia ecologically.

According to Soviet ecologists, 80–90 per cent of the land in Siberia and the north[7] should have been reserved for forests, swamps, steppes, and other natural landscapes, if one wanted to prevent irreversible disruption of the region's ecological balance. But the scientists had no illusions. One of the foremost said: 'But you probably know as well as I what we can really expect. First, the fish and the animals disappear, then come the forest fires and insects, blockages in the rivers, fellings, and finally, slides, mud torrents, floods and so forth' (quoted in ibid: 118).

And all of this did happen. Komarov gave many details, from which I am quoting only a small selection. For Komarov, there was no uglier and sadder sight than the streets and surroundings of the townships in Siberia and the north. Within a radius of several kilometres, there were no trees and no natural landscape any more. It was, instead, a dirty, swampy barren place, in which heaps of ash, slag, bottles, rusty car wrecks and wagons, and sundry trash lay around. In such townships, a lot more fuel per person was burnt than elsewhere. The combination of exhaust gases, fog and the low temperatures of the long winter produced frequent smogs with their well-known effects. The dirty, polluted snow and melt water caused the decay of moss and lichen cover, which in turn caused the frozen ground to become swampy. The forests were receding to the south, not only because of logging, but also because of this destruction of moss and lichen. Destruction of the forests in turn caused the velocity of the winds to rise and the temperature to fall (ibid: 114–15).

In the east, where the Baikal–Amur railway line was being built, many mountains were being deforested and no reforestation was taking place. Foresters were complaining that in the whole north-east of Siberia, less than 20 per cent of the logged forests were being reforested (ibid: 118). In western Siberia, where oil and natural gas were being extracted in a carelessly rapid process, there were numerous oil spills. Western Siberia is rich in water; in that region there are more swamps and lakes than solid ground. In the low temperatures of the north, oil slicks float on the surface for years and become no smaller. The oil industry there did not care about the environment and even discharged water pumped out of oil wells into lakes and rivers. This was a great danger for the whole environment of western Siberia (ibid: 127).

Waste disposal was and is a big problem in Siberia because of the frost and low temperatures. The everyday waste water does not get filtered. Decomposition of waste takes 20–30 times longer than elsewhere in the country. In spring, all waste reaches the rivers. And whereas a normal river can purify itself from a permissible dose of organic waste after flowing 200–300 km, for a river of the north even 1,500 km is not enough (ibid: 115–16).

It is not as if there was no public protest. Many scientists, journalists, citizens and interest groups did protest against many projects in the ways available and permitted in the country. The leaders also wanted to prevent the degradation of their country's environment. But this sincere desire was in contradiction to another much stronger desire: to catch up with the USA. The extra pure water of Lake Baikal, for example, was needed to produce 'super super cellulose cord' essential for aircraft tyres. Towards the end, environmentally harmful economic policies became vital for the very survival of the 'socialist' system. Export of oil, minerals, timber, and so on had to continue, and the growing consumption demands of the population had to be at least partly satisfied in order to avert a political crisis like the one that arose in Poland in 1980–81.

From reports published in the more liberal *perestroika* and *glasnost* period, it appears that neither Goldman nor Komarov had exaggerated. Some of these later reports, however, claimed that in the meantime the situation had improved to some extent. It was reported that in 1985 the Soviet Academy of Sciences received a special prize from the United Nations Environment Programme (UNEP) for saving Lake Baikal (Rosenbladt 1986: 149–50) and Michael Lemeschew wrote that about 80 per cent of the harmful substances in exhaust gases were being intercepted and detoxified, as a result of which the air in Moscow and some other cities had become noticeably better (Lemeschew 1988: 328).

But Sabine Rosenbladt reported from the large industrial city of Novokusnezk – where Mayakovsky had foreseen a garden blooming – that in 1986 the various enterprises there were still emitting 3,000 tons of dust and harmful gases every day, more than one million tons a year. That was the situation after the emissions had been reduced by 100,000 tons per year over the previous 10 years. Related diseases that officials reported were slow growth of bones in children, pathological changes of respiratory organs, chronic bronchitis, genetic defects, and others. The death rate was reported to be increasing (Rosenbladt 1986: 144–5, 150). The tenor of this report was, however, that ecological consciousness was growing and that the leaders were already doing more about the problem. Another spoke of 'some positive results' since the mid-1970s (Lemeschew 1988: 329).

Rosenbladt sounded excessively, and Lemeschew moderately, optimistic. But Alexei Yablokov, senior biologist and ecologist, and Gorbachev's chief ecological adviser in the second half of the 1980s, was very cautious. He showed that even in 1987–88, there was no substantial improvement. He wrote about 'ever faster extinction of whole animal and plant species' and about (in this connection) 'alarming news' from Lake Baikal. He reported that in 1987 the air of 104 big cities contained harmful substances more than ten times the maximum permissible concentration (MPC). As regards the rivers, he wrote that in every seventh one, the concentration of pollutants exceeded 10 times MPC. In the case of some lakes and reservoirs, it was perilous to eat their fish. In many areas of Central Asia, the mortality rate rose suddenly due to deterioration in the quality of drinking water. In general, the statistical life expectancy had not risen since 1971. In 1985, it was lower than in 1958.

As for the state of the land, Yablokov wrote that even in the second half of the 1980s more land was being lost than was being made useful through amelioration. In some parts of the country, the area of semi-desert was increasing at the rate of 10 per cent per year. The content of humus in soil, the basis of natural fertility, was decreasing (Jablokow [Yablokov] 1988: 307–10).

These reports are somewhat contradictory. The explanation seems to be that the earlier situation was very bad, and that even after spending a lot of money and achieving some 'positive results' in the previous decade (1975–85) the situation in the mid-1980s remained bad. An additional explanation that comes to mind is that economic growth, despite having slowed down considerably since the mid-1970s, nullified the effects of all the money spent on environmental protection.

ECONOMIC COSTS AND LOSSES RELATED TO ECOLOGICAL DEGRADATION

A real attempt by the USSR to pursue what is usually believed to be an ecologically sound policy of economic development would have given rise to costs. How much? Writing in 1970–71, Barry Commoner estimated that in the USA, for the same (or a similar) purpose, 'something like one half of the post-war productive enterprises' would have had to have been replaced by 'ecologically sounder ones'. It would have cost, at 1958 prices, 'about 600 billion dollars'. To this estimate one would have had to add the cost of efforts to restore damaged sectors of the ecosystem, which would have ranged 'in the area of hundreds of billion dollars'. This meant that most of the nation's resources for capital investment would have been needed for the task of 'ecological reconstruction for at least a generation' (Commoner 1971: 285).

Commoner's estimates were not exact for the USA of the 1970s, and cannot be applied without problem to the USSR of those days. But they give an idea of the enormity of the costs. Commoner spoke of replacing half of the enterprises. That was and is out of the question for the USSR. Perhaps what Petrovsky had asserted in 1968 – that it was forbidden to put industrial projects into operation if the purification installations had not been completed – was within the range of possibility. Even that was not done in the USSR. The investible funds were urgently needed elsewhere. It was a matter of choices and priorities.

The economic costs we are talking about should be classified in three categories: costs of preventing ecological degradation; costs of reversing (repairing or cleaning up) ecological degradation that had already taken place; and economic losses caused by ecological degradation, such as production losses. People suffering from diseases caused by ecological degradation give rise to costs in the second category, because a society must give medical treatment to such people, and in the third, because their productive capacities cannot be utilised. I have not found the (estimated) data neatly classified in terms of the above three categories. But for the purpose of this book, only a rough idea of the costs involved is necessary.

Basing himself on Soviet data available in the late 1960s, Goldman calculated in the early 1970s that 'it might cost 17–20 billion roubles just to improve the quality of the country's water supply and sewage processing'. But he also thought that this estimate might be too low (Goldman 1972: 119). He found it difficult to get data on the costs of eliminating air pollution in the USSR. He found one calculation made in 1969, according to which the Russian Republic spent 155.6 million roubles on air pollution control during the years 1959–67. This figure for by far the greater part of the USSR was less than 10 per cent of US expenditure on air pollution control over the same eight-year period, and American specialists were not at all satisfied with their national figure (ibid: 150).

Let me cite a few concrete examples of the costs involved: (1) Lake Sevan in Armenia – one of the highest lakes in the world and considered to be a natural wonder – was in the process of eutrophication. Water from other areas was diverted into it in order to save it. This measure was estimated to have cost US$100 million (ibid: 66). (2) In 1970, the cost of keeping Lake Baikal partially clean by means of waste-water treatment plant was about 20,000 roubles a day (ibid: 199). (3) It was estimated that over the years 1968–80, 430 million roubles would have to be spent for Moscow's sewage treatment plant (ibid: 103). (4) Goldman found a report stating that just one dust collector for an open hearth furnace cost 1.5 million roubles in 1968. Therefore the total figure of 155.6 million roubles that the Russian Republic spent on air pollution control in the period 1959–67 was enough for only 100 such dust collectors (ibid: 150).

Of course, there were also technical problems. Complaints about lack of research institutes and production facilities for pollution control equipment were heard from almost all of the republics (ibid: 147–8). But shortage of funds was the main cause of lack of research institutes. Goldman wrote in the same context: '... actual installations of pollution control equipment are far below the number of requests for help received from factories that pollute. These requests in turn are far below the number of complaints and demands for action issued by government authorities' (ibid: 148). It is noteworthy that complaints and demands for action were issued by government authorities, although there was no ecology movement in the USSR comparable to that in the West. It is therefore clear that as far as ecological consciousness and intentions were concerned, the USSR was not lagging behind, simply funds were lacking. Goldman commented in a somewhat different context : 'If enough money is available, solutions can always be found. Nature takes its revenge, however, because invariably these new corrections tend to generate their own negative consequences which require yet additional rectifications' (ibid: 238). I shall return to this theme in chapter 4. For the present, let us consider Komarov's report on this question 6–7 years later.

Komarov quoted Lemeschew, who, in the mid-1970s, opined that the Soviet economy could not afford to spend 15–20 per cent of every factory's production costs to protect the environment. According to him, fitting filters to the exhaust pipes of motor vehicles, which would improve the air of the cities, was an 'inadmissible luxury' (cf. Komarov 1980: 29). A viable proposal to save Lake Baikal, involving the transfer of effluents from the paper and pulp factories through a pipeline over the mountains to the Irkut River, could not be accepted because of its cost (US$40 million), which would have made the whole enterprise unprofitable (ibid: 9; Goldman 1972: 202).

Komarov wrote a noteworthy but unclear sentence about the costs of protecting the environment. Referring to the estimates of unnamed Soviet economists, he reported that in the mid-1960s, Soviet society could compensate for one rouble of damage or loss due to pollution by spending 50 kopecks on environmental protection. In 1977, for the same purpose, 1.5–1.7 roubles were needed; in the early 1980s, 3–3.5 roubles (Komarov 1980: 136). If these estimates are correct, then the logical conclusion must be that, after a certain point, trying to compensate economic loss or damage caused by environmental disruption by spending money to reverse or rectify it becomes meaningless. It was, therefore, at least *economically* rational if, after 1977, the Soviet authorities had done nothing to rectify environmental disruption.

Here, two objections are possible. Firstly, damage to citizens' health not only causes economic losses, but is in itself a loss in the sense of human suffering. In this matter, cost should not be any consideration. But then there would have been fewer consumer goods to enjoy. Secondly, one may object that the correct environment policy is to prevent disruption rather than to rectify it subsequently, in which case the costs would not be too high. That is right. But even fitting the exhaust pipes of motor vehicles with filters, actually a preventive measure, was for Lemeschew an 'inadmissible luxury'. And the production of such filters also contributes to environmental degradation.

Let us come back to the losses. The authors of the two forecasts quoted by Komarov (*Our Nature in 1980* and *Our Nature in 1990*) roughly quantified economic losses in the narrow sense: by 1980, total annual losses due to air and water pollution would amount to 20 billion roubles. By 1990, they estimated, the figure would rise to 45 billion roubles. But Komarov, who had access to secret materials, wrote that according to some other documents, total losses in 1980 would probably amount to 50–60 billion roubles and in 1990 to 120 billion roubles (ibid: 25–6).

From a later source, we have an official figure. In 1984, the vice-chairperson of the Environment Committee of the Supreme Soviet quoted estimates according to which annual losses due to air and water pollution amounted to 2.5 per cent of Soviet GNP. If one added the cost of measures that had become necessary to protect the soil, then the losses came to 4.5 per cent (cf. Rosenbladt 1986: 148–9).

Even if we accept this official figure, and if we consider how problematical the concept GNP is, then it seems that, all things considered, no real improvement in the economy was taking place around this time. For in the five-year period 1976–80, the USSR's GNP rose (again, according to official data) by only 21 per cent (Aganbegyan 1988: 10), an average of 4.2 per cent per annum. This alone would be sufficient to explain the stagnation in the economy.

Komarov reported that half of the losses due to ecological degradation occurred in the health sector. From the two forecasts, he quoted a few telling details of ill-health due to air pollution: in the ten years before he wrote his book, the incidence of lung cancer doubled; and the incidence of genetic damage in new-born babies grew by 5–6 per cent each year. At the time Komarov was writing, 7–8 per cent of the population had genetic defects, and it was feared that, by 1990, the figure could rise to 15 per cent (Komarov 1980: 25).

Let me now quote a few data on the situation towards the end of the 1980s, as published in 1991 by the Worldwatch Institute. (1) 'The health costs alone of pollution in the Soviet Union were reported to be 190 billion roubles in 1987 (US$330 billion at the October 1990 official rate) or 11 per cent of the nation's estimated GNP.' (2) 'Production losses attributable to soil erosion cost the Soviet economy 18–20 billion roubles (US$31–5 billion) annually.' (3) 'As immediate measures to reduce pollution, the government of the USSR shut down 240 factories in 1989. That could lead to a backlash: many plants producing essential medicines and scarce paper have been closed.' (4) 'According to one estimate, the Soviet Union must spend 100 billion roubles (US$174 billion) initially and then 10 billion roubles annually just to cut air pollution to the accepted limits' (French 1991: 94, 101, 106, 108).

In the passage quoted on pp. 35–6, Marx accused 'capitalistic agriculture' and 'capitalist production' of 'sapping the original sources of all wealth – the soil and the labourer'. It seems now that also 'socialist' production sapped the soil and the labourers of the USSR. Marx himself unknowingly pointed at the most important similarity between the two systems when he wrote in the same passage: 'The more a country starts its development on the foundation of modern industry, … the more rapid is this process of destruction'. Both the USA and the USSR did exactly that.

What percentage of its GNP was the USSR spending for the protection of its environment, and how much would be sufficient for the purpose? According to Yablokov, the two figures were 1 per cent and 5–6 per cent respectively (Jablokow 1988: 312). But the USSR simply could not spend that much on environmental protection and still retain its status as a super-power. Moreover, it is doubtful whether allocating more funds for the protection of the environment is at all meaningful if the extra funds are to come from economic growth, and are invested in technological devices. Such a strategy is self-defeating, because economic growth leads to further ecological degradation.

Perestroika-enthusiasts believed that economic growth was possible without worsening ecological degradation and simultaneously with diminishing production of raw materials and intermediate goods. Lemeschew's solution, for example, was 'that one is able to cultivate less land of better quality with a smaller number of first-class tractors and to produce, by fulfilling all the requirements of agro-technology, larger quantities of cheaper agricultural goods of better quality' (Lemeschew 1988: 342–3). Of course, that would be the dream solution. That is called nowadays 'sustainable growth'. But I have strong doubts about its feasibility. Lemeschew himself said elsewhere in his paper that to develop new waste-reducing and waste-free production technologies would require additional expenditures that would consequently make the products more costly (ibid: 335). And where will better quality land come from? Quality of land is a given factor. It can be improved to some extent, but not without effort and expenditure. I shall address these questions in chapter 4.

LACK OF COLONIES

In connection with the resource problem, Aganbegyan wrote:

> the USSR is perhaps the one country in the world the greater part of whose fuel and raw materials requirements is met through domestic production, not imports. ... There is neither the opportunity nor the need to behave like the developed capitalist countries in meeting their fuel and raw materials supplies. If one of these countries needs additional quantities of coal or non-ferrous metals, it purchases them as a rule from underdeveloped countries, investing capital there when necessary, creating its own companies there. We cannot behave in this way.
>
> Our approach relates ... to a certain extent to labour also. Any shortage of scientists or highly qualified specialists in the USA has been resolved through 'recruiting' candidates from other countries through the attraction of higher salaries. Turkish and Yugoslav workers were attracted to West Germany when additional unskilled labour was needed. When this need ended, foreign workers were immediately discriminated against. Such a course also is unacceptable to us. (Aganbegyan 1988: 9)

It seems to me that Aganbegyan here makes a virtue of an inability. As far as fuel and raw materials were concerned, the USSR, of course, had enough. But the costs of production were, as we have seen, rising rapidly. Economic rationality should have required them to opt for cheaper imports. But they could not do that to the extent needed. Aganbegyan also admits that there was no 'opportunity' to behave like the developed capitalist countries in this respect. As for personnel, scientists and highly qualified specialists from other countries never thought of taking a job in the USSR on account of the salaries being so low. Perhaps the Soviets had sufficient experts. But as regards workers, the USSR admitted to a shortage (ibid: 7–8). It is quite well known that the GDR recruited large numbers of workers from Vietnam, Angola, Mozambique, and other places; it is not known whether, or to what extent, the USSR did the same.

The general shortage of workers was due mainly to the low level of labour productivity. We have seen that in the Soviet mining industry, labour intensity was growing. But also in the economy as a whole, the rate of growth of labour productivity was falling or stagnating (ibid: 68). The main cause of this was the low level of automation, and the main cause of that was the low level of funds available for investment in automation. Low labour productivity is generally the cause of low wages. The wages in the GDR were, of course, attractive enough for workers from Vietnam, Angola and Mozambique, but not for Turkish and Yugoslav workers. The same possibly applied to Soviet wages; the harsh climatic conditions were certainly an additional strong disincentive.

The main reason that the USSR did not have the opportunity to behave like the developed capitalist countries in regard to raw materials and workers was its chronically weak position in the world market. Its technology and high-value industrial products were neither highly regarded nor available in sufficient quantities (after meeting domestic demand and the demands of other 'socialist' countries) for export to hard-currency countries or to countries that could pay in hard currency. As a result, its main exports were low-value fuel and other raw materials.

Let us consider some important data on Soviet foreign trade in the mid-1980s. The USSR's share in world industrial production was roughly 20 per cent, but its share of world trade was only 5 per cent (ibid: 148). Only 12 per cent of its national income came from exports, while many capitalist and 'socialist' countries of Europe derived as much as 40–50 per cent from this source[8] (ibid: 141). There is, however, nothing negative in these data. The USSR could satisfy many more of its needs through its own production than the other countries of the world. But the composition of its foreign trade showed negative traits. Fuel, electricity, and other raw materials, such as ores, timber, and so on, accounted for 62 per cent of its exports, while machines, equipment and vehicles accounted for only 14 per cent (ibid). The fall in world market prices of already low-valued fuels and raw materials had negative effects on the Soviet economy similar to those on the economies of underdeveloped countries.

The USSR's low share in world trade was also partly a result of history and politics. In order to achieve rapid industrialisation and to protect the 'socialist' system against foreign economic competition and military threat, it was necessary to pursue a policy of relative autarky. However, in two very important respects, the conditions for economic development in the second half of the 20th century were very different from those in the first half. Firstly, the number of technologies that had to be mastered had grown enormously, and all technologies had become extremely complicated, so that they required a much larger and much more varied resource base. Moreover, technological progress in general had become very rapid. Secondly, the amount and especially the kinds of consumer goods necessary to satisfy the population of any country had grown enormously. No one country, not even a group of countries, was in a position to produce all of them, nor in a position to produce the intermediate products necessary for the final products – at least not in an economically viable way.

For these two reasons, it became extremely difficult and expensive – almost impossible – for an economically backward country to bridge the large technological gap by its own efforts. The USSR failed in this respect, despite its achievements in space and armaments technologies and in spite of the huge natural resources its territory contained.

But its conception of socialism as well as its other ambitions – becoming a super-power, catching up with the USA in standard of living, higher national prestige – made it imperative for the USSR to close the technological gap, and also to import many more goods and technologies than in the 1920s and 1930s. That was no longer possible with the old policy of relative autarky. Integration into the world market, and economic co-operation and competition with the other industrial nations became absolutely necessary.

In this connection, an important difference between the leading Western industrial countries and the 'socialist' countries of Europe must be borne in mind. In the former, the process of primitive capital accumulation had been greatly facilitated by the conquest and exploitation of the colonies – exploitation of both natural and labour resources, including slave labour. A large part of the higher standard of living in these countries was directly made possible through this exploitation.

Even after the colonies became politically independent, economic exploitation continued by other means. It has even been further intensified, owing to the higher development of productive forces.[9] This is known as neo-colonialism. The basis of neo-colonialism is the difference in command over the Earth's resources, over finance capital, over the accumulated scientific, technological and managerial know-how of humanity, and over commercial contacts worldwide. This difference arose largely from colonialism. The world's few rich capitalist countries, and their associate countries, have by far the greater part of these factors of prosperity at their command. They defend their command over these factors by all means – financial, commercial, political and military. They also have the whole underdeveloped world's cheap labour power at their disposal.[10]

The 'socialist' countries of Europe did not have these – neither colonies nor neo-colonies in the above sense. Logically, all countries of the world could not and cannot have colonies and neo-colonies. There is also a limit to colonies in this finite world. Czarist Russia was, of course, a colonial power. But it could not exploit its European territories – the Ukraine, Belorussia etc. – as, say, Britain exploited India. Neither Siberia nor arid central Asia were very lucrative colonies – not with the technologies of those days. We have already seen that even with 20th century technologies, it has not been easy to exploit the resources of Siberia. And by the time Russia could begin its industrialisation, it had become a member, albeit the leading one, of the 'socialist' USSR. The people living in the former Czarist colonies had become equal citizens of the same 'socialist' state. They could not be exploited. On the contrary, 'socialists' had the ambition to develop them to the same level as the people living in Moscow and Leningrad. There was no cheap labour power there to be exploited and no one to rob.

As for labour costs in Siberia, workers from other parts of the country had to be enticed to go there with three times average wages and many other extra facilities. Even then, nobody really wanted to settle there. Most people worked there for only three years, the minimum contracted period, took their premium – a car – and left (Komarov 1980: 119, 127–8).

For the Western industrial countries, however, the costs of importing the resources and primary products of the underdeveloped countries – their former colonies and present-day neo-colonies – have been continuously decreasing. Thus between 1961 and 1964, western Europeans could get, for one Swiss watch, 7.5 kg of Tanzanian coffee. Ten years later, between 1971 and 1974, they could get, in exchange for a similar Swiss watch, 14.2 kg of Tanzanian coffee (Strahm 1981: 46). In 1985, western Europeans could get, for a West German truck, 7.6 tons of cocoa from the Ivory Coast. In 1990, they could get 29 tons (*Frankfurter Rundschau*, 24 July 1991).

Two main points emerge from this difference. Firstly, it partly explains the rising standard of living in the West, and, conversely, it partly explains why the standard of living in the industrialised 'socialist' countries lagged behind that of the industrialised, capitalist West. Secondly, it explains at least a part of the economic growth in the West. In December 1985, *The Economist*, the journal of British capital, wrote:

> Just as yesterday's dearer oil twice brought stagflation[11] to the rich countries, so today's cheaper raw materials can now help them kill inflation and boost economic growth. ...
>
> Most of the rich countries have not yet understood ... how much of it is

> a gift from poor countries. In the past 12 months ... the price of metals has dropped by 15%, that of oil by 5.5%. These declines ... mean that consumers are now paying about $65 billion a year less for the same amount of raw materials than they did 12 months ago. (*The Economist*, 6 December 1985: p. 13)

The USSR and the other 'socialist' countries could hardly get anything similar to this 'poor man's gift' to the Western rich.

Of course, in the 1970s the Chinese and their followers accused the USSR of being imperialistic in its trade relations with underdeveloped countries. But the accusation was not convincing. They quoted some facts and figures, but on closer examination, these proved inadequate to justify the accusation. This is not the place to examine the matter in detail, but one or two examples should be given.

It was said that the USSR was charging a higher price for its exports to underdeveloped countries than for similar exports to western industrialised countries. For instance, in 1974, it was demanding for a car exported to underdeveloped countries 1,569 roubles, but for a similar car exported to western Europe only 805 roubles (N.K. Chandra, quoted in Bogen 1987: 24). At first sight, this appears to be exploitation of underdeveloped countries. But actually, it proves only that the USSR was selling its cars in western Europe at a dumping price, perhaps at a loss. In any case, the underdeveloped countries were surely not paying a higher price than they would have paid for a similar Italian or French car. The low price for western Europe reflected only the USSR's eagerness to earn hard currency. This reason did not exist in the case of underdeveloped countries, for the USSR's trade with them was mostly on a barter basis.

It was found that in the period 1960–69, of the 12 leading commodities that India exported to the USSR, only one, unmanufactured tobacco, consistently received a price below the world market price (six consistently received higher than world market prices) (Asha Datar, quoted in Szymanski 1979: 162). But there are various qualities of unmanufactured tobacco; the lower price paid by the Soviets could easily be explained by their having imported lower quality tobacco (ibid).

What the Chinese and their followers could prove, however, was that the USSR, in its trade relations with the underdeveloped countries, was not behaving in a self-sacrificing manner except in the case of Cuba. It was trying to get some benefit out of this trade. But that is the purpose of any trade. One can only speak of exploitation when one side can derive a higher rate of benefit (profit) than the other side because the latter finds itself in a situation of direct or indirect compulsion.

It can justifiably be argued that the world market prices of most export commodities from underdeveloped countries are exploitatively low, and that the world market prices that they pay for their imports from advanced industrial countries are for them exploitatively high. So, by simply sticking to world market prices, the USSR was exploiting underdeveloped countries.

Even if one accepts this argument, why could not the exploitation of underdeveloped countries help the USSR to overcome its economic crisis? The answer is very simple. The volume of the USSR's trade with the underdeveloped countries was too low. The trade between the 'socialist' countries of Europe and the underdeveloped countries amounted in 1982 to only 2 per cent of world trade. The USSR's share in this 2 per cent was 58 per cent (Bogen 1987: 13). In 1986, the USSR's trade with such countries had a total volume of only 14 billion roubles (Aganbegyan 1988: 153). The total profit from this was too little to make any substantial contribution towards helping the USSR to overcome its crisis.

Another important point must be noted. Until the early 1990s, there was a net annual flow of money from the underdeveloped countries to the industrialised countries resulting from the external indebtedness of the former. In 1989, this amounted to US$42 billion (Deutsches Institut für Entwicklungspolitik 1991: 1). In 1992, it was reported to be roughly US$50 billion. Another report, which includes profit remittances of transnational corporations (TNCs) and flight capital from underdeveloped countries to the industrial countries of the West, estimated the figure to be US$120 billion (*Der Spiegel*, 18 May 1992). It should be noted that these are net figures, arrived at after deducting money flowing from the industrialised West to the underdeveloped countries: commercial credits, 'development aids' – which are in any case mostly credits – and private and corporate investments.

Since the mid-1990s, the flow has been reversed owing to massive direct private and corporate investments from the industrial countries into *a few* underdeveloped countries (cf. Wahl 1997). But credits and investments are not final flows. Credits have to be paid back with interest, and investments generate profits that finally flow out in one form or another. Apart from the fact that such private credits and short-term investments have a volatile and negative aspect, as evidenced by the sudden financial and economic crises of late 1997 in east and south-east Asia, the debt crisis of the underdeveloped countries as a whole has not been mitigated by such inward flows. On the contrary: according to a recent report, it has worsened. From World Bank data it appears that in 1997 the total liabilities of the underdeveloped countries have increased by about 10 per cent (*Frankfurter Rundschau*, 20 January 1998).

The USSR and a few other 'socialist' countries also gave credits in the name of development aid. But they were given to friendly states at very low rates of interest. There were hardly any 'socialist' companies in underdeveloped countries, although there were quite a few in advanced capitalist countries (Levinson 1980: 81–2).

In the 1960s and 1970s, western capital entered the USSR and other 'socialist' countries. It began as large credits given to finance enterprises to be built with the help of western TNCs,[12] for example, the large Fiat car factory in Togliattigrad. (A large part of the American credits were for wheat purchases.) The idea was to import advanced western technology in this way. But by 1976, the USSR had to admit that external indebtedness had become a problem for itself and some other 'socialist' countries (cf. ibid: 25). A solution was sought by re-exporting some products of the enterprises built in 'socialist' countries with western technology and credits. This was called co-production.

Here a certain similarity was evident between the European 'socialist' countries and the underdeveloped countries in their economic relations to the West. The difference was in the degree of dependence. Foreign indebtedness was there. Exploitation of cheap 'socialist' labour by western TNCs was there. In the 1980s, the GDR even became a dumping ground for all kinds of waste from West Germany.

In any case, it was clear that the USSR and the other 'socialist' countries of Europe could not solve their economic problems, nor even close the technology gap to any significant extent, by integrating themselves in the world market. Their share in world trade and in the profits therefrom was too low – a situation similar to that of most underdeveloped countries. So they could not catch up with the USA, Britain or the FRG. They bumped up against the limits to growth.

Having failed to build 'socialism', they became eager to do what they felt to be the next best thing: to become junior partners of the West. Their concrete business contacts with western capitalists and their moral degeneration, which I shall describe in the next chapter, prepared them for the transition to capitalism.

NOTES

1. It must be said here that not all 'socialist' regimes in Eastern Europe were as terroristic as the Stalinist one in the USSR.

2. In fact, Gorbachev did exactly that in the USSR, in the *perestroika* and *glasnost* period. Elections took place in which non-communists could become

deputies by defeating communist candidates, although the dominance of the Communist Party as the only party and the ruling party was guaranteed. But the whole process ultimately led to a total breakdown because the economic basis was very weak and became weaker exactly because of *perestroika* and *glasnost*. Before Gorbachev, Khrushchev had initiated some political liberalisation which Brezhnev did not fully revoke.

3. It was a pseudonym. Komarov succeeded in getting the book published in the West apart from circulating it in the underground dissident scene.

4. I have used the term 'degradation' to mean both pollution and all those changes (e.g., clear-felling) that disturb, disrupt, or destroy the ecological balance but cannot be called pollution.

5. Aganbegyan obviously means annual production. Otherwise the investment won't make sense.

6. This argument is, however, no longer valid. The polluter can pass off his costs to other societies in other states. The validity of the argument can be restored only if the whole world becomes one state and the whole of humanity one society. Moreover, the present generations can pass off a part of their costs onto future generations in the same state, even if the state is a socialist one.

7. From the term 'Siberia and the north', Komarov excluded those parts of Siberia that have a temperate climate.

8. It should be noted here that these figures include exports from the USSR and other 'socialist' countries to each other. They, firstly, did not bring in any hard currency, and, secondly, did not prove that products of the 'socialist' economies were attractive in the world market. To import modern technology and food products from the West, however, exports to hard currency areas were necessary. Such exports were, obviously, much less than the total figure.

9. Another factor that contributed to this intensification was the continuous growth of population, which made ever more cheap labour available to the exploiters. It must, however, be added here that also local and national exploiters profited from these two factors of intensification.

10. It must be mentioned here that since a few years ago some leftists and former leftists are denying the fact of neo-colonial exploitation of the underdeveloped countries. But I do not find their arguments convincing.

11. The term 'stagflation' was coined in the mid-1970s to denote the unusual situation in which the economy suffered from both inflation and stagnation, that is, zero or very low growth.

12. In those days, Khrushchev was pursuing a policy of peaceful coexistence and peaceful socio-economic competition of the two systems.

CHAPTER 3

Why the Soviet Model of 'Socialism' Failed (2): The new class and the moral degeneration of 'socialist' society

In chapter 2, I presented the most important cause of the failure of 'socialism'. In this chapter, I shall present and discuss the second major cause, namely the origin of a new exploiting class among the communists and the moral degeneration not only of the communists but of the whole 'socialist' society. But let us first look at the ideal and then see what became of it under the leadership of the Communist Party.

THE IDEAL

Classless society

The most important argument for the view that the 'socialist' societies were not really socialist is that socialism is unthinkable without democracy, and since the 'socialist' societies were not democratic according to the usual interpretation of the concept, these societies were not socialist. The basic counter-argument of communists has been that no society which is divided into classes on the basis of differences in wealth and income and is dominated by a minority class which exploits the weaker classes deserves to be called democratic. The most important criterion of democracy has been for them the absence of exploitation, and since the 'socialist' societies were, according to them, free from exploitation, they were basically democratic.

Of course, it can be argued that even a classless society can be undemocratic. But if it can be shown that the 'socialist' societies were free from exploiting and exploited classes, then the view that they were democratic, and hence really socialist, becomes at least more plausible. As early as 1936, it was claimed that in the USSR 'all the exploiting classes' had been 'eliminated', that 'the exploitation of man by man' had been 'abolished' and that 'the complete victory of the socialist system in all spheres of the national economy' had become 'a fact'

(Stalin 1976: 799–800). Earlier, the society was considered to be a dictatorship of the proletariat, which, in the understanding of the communists, was not only 'socialist', but also democratic or at least close to that, because it was dominated by the working people, the producing classes, who were the majority of the people.

According to communists, Soviet society, from the second half of the 1930s, was made up of two classes – workers and collectivised peasants – and a stratum of intelligentsia. But there was allegedly no antagonism between them; they lived and worked in friendly collaboration, and they earned according to the work done.

The new man[1]

From the very beginning of the revolutionary regime, a conception of the ideal human – called the new man – was propagated and cultivated as a corollary and prerequisite of the ideal of 'socialism'. It was thought that just nationalising or collectivising the economy was not enough to build a 'socialist' society. Andrei Sinyavsky wrote in his critical book on Soviet civilisation in 1989: 'The idea of the New Man is the cornerstone of Soviet civilisation. Without him, building up socialism in a backward country is impossible. From the very beginning, the idea of a radical and inner change and transformation of the human being was connected with the revolution' (Sinyavsky 1990: 115). He then quoted the famous Soviet poet Mayakovsky, who wrote the following verse in 1918:

> 'We fired for a day or two
> Then we thought:
> We'll clean the old man's clock.
> That's what!
> Reversing one's jacket,
> That's not enough comrades!
> Go on, turn yourselves inside out!'
>
> (quoted in ibid)

Bukharin, a leader and theoretician of the Party, wrote in 1922 that the real task of the revolution was 'to alter people's actual psychology'. In 1928, he wrote: 'In our system of scientific planning, one of the first priorities is the question of the systematic preparation of new men, the builders of socialism' (quoted in ibid).

Sinyavsky wrote that in the course of a few years the term became identical with the term 'communist'. The qualities of the new man were deemed to be as follows: Firstly, 'boundless devotion to the supreme goal, namely the building of an ideal society on earth'. Secondly, 'the ability to shift resolutely from words to deeds. The new man, forever remaking the world in an image closer to his ideal, is not a dreamer but a doer.' And thirdly, 'the new man has the habit of representing the masses or the class, which effects its own ideal through him'. He/she is never an individualist: 'the new man toils not for himself but for the great common cause'. That is why 'the most egregious sin, from the perspective of the new man, is egoism or individualism, the desire to live for oneself as opposed to the common good' (ibid: 116)

According to this theory, 'bourgeois' was not simply a class category, but above all a psychic phenomenon. Bourgeois relics were believed to continue to exist in people's consciousness even after the liquidation of private property and the bourgeoisie as a class. These relics were individualism, indolence, profligacy, addiction to personal profit, and so on. The new man was therefore proud of not having anything personal, of having renounced his or her interests in favour of the common weal.

Sidney and Beatrice Webb gave an account of communist ethics prevailing in the first half of the 1930s. In 1932, they asked a 'most influential and most widely respected Bolshevik leader', 'an embodiment of the conscience of the Communist Party' the following question: 'What is the criterion of good or bad in the conduct of a member of the Communist Party?' His answer was, substantially: 'Whatever conduced to the building up of the classless society was good, and whatever impeded it was bad' (Webb 1944: 838). In this context, the Webbs observed: 'It is, indeed, a fundamental principle of communist ethics that every individual should actively strive to bring about a condition of social equality. He must insist on the complete abolition of privilege ...' (ibid: 857). The Webbs summarised the ideals that Bolshevist teachers and writers were propagating in their perpetual campaign of education as follows:

> Not the perfecting of one's own soul or self, but the service of others, and the advancement of the community, constitutes virtue. No one is deemed to be good unless he does what he can for his fellow-men. He is not judged by his works, for his works may be unsuccessful from no fault of his own, but by the *motives* and *incentives* that govern his actions. Even if his works are socially useful and successful, if he is a 'careerist' or a 'self-seeker', he is not a good man. (ibid: 839)

On the theme of personal acquisitiveness, the Webbs wrote: 'There is, in the USSR, a widespread and persistent discouragement of personal acquisitiveness ... What is "not done" under Soviet communism is the seeking of personal riches' (ibid: 853).

The Webbs reported that pecuniary saving by the individual had ceased to be a recognised virtue. Although the state paid interest and dividends on savings and bonds, trade unions tried to motivate workers to take, as a patriotic duty, a share in internal loans, which yielded no interest but only lottery prizes on the drawn bonds. This was universally regarded as a sacrifice of consumption. Many prize-winners even omitted to claim their prizes (ibid: 855–6). The Webbs added that Party members were generally held to a higher standard of personal conduct than ordinary citizens. They had 'voluntarily pledged themselves to poverty, to the extent of never accepting for themselves any larger salary or wages than the common maximum laid down by the Party rule …' (ibid: 856). It was an experiment unprecedented in world history. Another contemporary Western observer, Ella Winter, commented: 'No society … has heretofore attempted to create its morality consciously. … The new man is planned as the new society is projected' (quoted in ibid: 842). Lunacharsky, People's Commissar for education, said in 1931: 'We must do all in our power to create a new man with a new psychology' (quoted in ibid.).

THE REALITY

It must have been difficult, especially for foreign observers like the Webbs, to know to what extent reality corresponded to ideals and principles. Critics of the Soviet communists asserted even in the early years of the regime that a new system of privileges, inequalities, and despotism was taking shape there. In 1919, Kautsky, a German social democratic leader, used the term 'new class' in this connection (cf. Kolakowski 1978: 163). In 1957, Milovan Djilas, until a few years earlier a leading Yugoslav communist, published his attack on communism under the title *The New Class*. And in the 1960s, many leftists outside the 'socialist' societies started saying that the USSR and the other 'socialist' societies of Europe were *no longer* socialist.

From the point of view of these critics, therefore, at least in the USSR and the Eastern European countries, 'socialism' had failed very early, albeit only because it could not defend itself against the attacks of the new bourgeoisie.

The class question

How big was the gap between the highest and the lowest income? What was the level of pay in each profession? Were income differentials growing or shrinking? Answers to such questions are of great importance in determining whether the 'socialist' societies were really socialist or had become class societies. In spite of doubts and difficulties with regard to data from the USSR, some general conclusions are possible. One can generally speak of a pendulum-like movement between egalitarianism and anti-egalitarianism.

In the first few years after the revolution, equality was the ideal. It was generally expected that there would no longer be any difference, as far as material rewards were concerned, between industrial and agricultural work and between manual and intellectual work. As far as communists were concerned, the ideal was also practised. The salary of a People's Commissar was comparable to a skilled worker's wage, and party members who earned salaries above a certain level had to hand over the surplus to the party (cf. Carr 1976: 117–18). (This rule was known as *partmax*.)

But despite these ideals, the policy actually pursued for the society at large was rather pragmatic. The new regime more or less continued the practice of differential wages inherited from the previous regime. The circumstances were simply not conducive to promoting motivation with moral incentives only. In this period, restoring production after the war was the most important task.

In the period of war communism (roughly between summer 1918 and spring 1921, which was also the period of the civil war), there was a policy to narrow the spread between unskilled and skilled workers that existed before 1914. In February 1919, the wage differential in the category 'workers' was fixed at 1:1.75. In higher-paid groups it was also 1:1.75. The ratio of minimum wage to highest salary for the best-qualified administrative personnel was 1:5 (ibid: 205).

In 1920, it was felt necessary to widen wage differentials. But money wages had become almost meaningless and payments in kind – rations – were increasingly replacing them. Of course, rations were differentiated according to status and occupation, but it was not possible to adjust rations to individual output. An attempt was made to introduce a bonus in kind, but it failed due to shortage of supplies. It was almost impossible to distribute anything in excess of the barest minimum rations. Carr sums up: 'The ultimate result of war communism in the field of labour policy was to leave no other incentives in operation except revolutionary enthusiasm and naked compulsion' (ibid: 220).

With the introduction of the New Economic Policy (NEP), private enterprise was allowed and the Soviet Union developed a mixed economy. So far as wages were concerned, the state's task was now merely to fix a minimum, actual wages being a negotiable matter. Egalitarianism was buried and wages were linked to productivity. The

NEP brought back inequality to what it had been before the revolution. Towards the end of the NEP, according to the calculations of Peter Wiles, inequality among workers and employees exceeded the degree of inequality in Great Britain in 1966. While in the USSR, in 1928, the decile ratio[2] was 1:3.8, in Great Britain in 1966 it was 1:2.7 (both after tax) (Wiles 1981: 22).

This unsocialistic aspect of the USSR obviously embarrassed some Bolshevik leaders. But Stalin put an end to that when he became the supreme leader. In 1931, he denounced egalitarianism in wages and salaries as 'alien and detrimental to socialist production' (quoted in *Encyclopaedia Americana* 1982: 401). He condemned the 'leftist practice of wage equalisation'. He justified this in the following words:

> We can no longer tolerate the situation where an iron-founder is paid the same as a cleaner and an engine driver no more than a copyist. The consequence of wage equalisation is that the unskilled worker lacks the incentive to become a skilled worker and is thus deprived of the prospect of advancement; as a result he feels himself a 'sojourner' in the factory, working only temporarily so as to earn a little and then going off to 'seek his fortune' elsewhere. (quoted in Dobb 1993: 461)

The whole Stalin era was one of extremely wide income differentials. For instance, the income differential between the upper categories of workers and the 'auxiliaries' (cleaners, messengers, etc.), who were ranked below the workers, reached 1:12 (Nove 1982: 212). Engineers, technicians and factory directors also gained from this trend. Shortly before the Second World War, the average earnings of workers were between 200 and 300 roubles, while top-ranking scientists sometimes earned as much as 2,500 roubles (Dobb 1993: 467), a ratio of between 1:12.5 and 1:8.33. Let us remember that in 1919 the ratio of the minimum wage to the highest salary was only 1:5.

In the post-war years, there was a general 'wage drift' in favour of workers on piece-rates, that is, the great majority in the category 'worker'. From the late 1950s, there was a steady trend towards reducing the extreme income differentials of the Stalin era. In 1956, steps were taken to establish (and later increase) a minimum wage (in the Stalin era there was apparently no minimum wage). The downward trend of income differentials continued into the second half of the 1960s. In 1966, the ratio of the lowest average to the highest average income was 1:2.28 (Szymanski 1979: 67).

Between 1965 and 1973, collective-farm peasants and state-farm workers, the two groups of working people most neglected in the Stalin era, gained the most – increases of 78 per cent and 62 per cent respectively (ibid: 52). In the late 1970s, income differentials were reduced further, mainly through successive rises in the minimum wage. In 1978, the differentials among workers ranged from 1:1.86 (in mining) to 1:1.58 (in the light and food industry). A foreman's pay was roughly equal to the average wage of workers, that is, lower than the income of highly skilled workers (Nove 1982: 213).

Of course, tariff wages and salaries do not tell the full story. Managers of enterprises got bonuses for fulfilment of plans of about 25–30 per cent of their salaries (ibid: 66), and perks such as free cars and summer houses. But workers also got bonuses and a social wage. Moreover, the pricing policy of the state had some equalising effect on real incomes: prices of basic necessities were below their value and those of luxury goods above theirs.

Szymanski showed that in 1973 the average annual income of American male self-employed professionals and salaried physicians and surgeons was 2.4–2.7 times the average annual wage of an American industrial worker. This was approximately the ratio of the incomes of highest-level Soviet enterprise managers, government officials, and state ministers to the average income of Soviet industrial workers. The former earned 2.7–4 times more than the latter (Szymanski 1979: 63–6). He concluded: 'This suggests that the leading stratum in the Soviet Union is, at least in income terms, similar to the US professional petty bourgeoisie, rather than to the American capitalist class (of either corporate managers or the multi-millionaire wealthy)' (ibid: 66). Szymanski examined a lot of data from the 1960s and early 1970s in order to find out whether, and if so to what extent, the stratum with incomes comparable to those of American professionals was forming into a distinct social class. His conclusions were as follows: first, the professional or scientific–technical intelligentsia showed a tendency to crystallise as a distinct social stratum separate from both the manual working class and the managerial–political intelligentsia. But the tendency for one or both of the two categories of intelligentsia to crystallise into a social *class* was very weak. Second, there was no 'elite privileged social stratum' with its own 'highly distinctive life-style, exclusive intermarriage patterns and virtual certainty of passing on its positions to its children ...'. Third, the Soviet power elite did not form a distinct social class comparable to the wealthy corporate owning and managerial class in the West.

He especially stressed the remarkable upward social mobility resulting from the state's efforts to equalise educational opportunities. The best results were achieved in the area of politics. In 1966, 40 per cent of the economic ministers of the Union government had manual working-class parents, 27 per cent had peasant parents, the parents of 15 per cent were low-level white-collar workers, and the parents of only 18 per cent belonged to the intelligentsia. In the same year, among the members of the Central Committee of the Party, 36 per cent had manual working-class parents, 47 per cent peasant parents and only 16 per cent non-manual (either intelligentsia or low-level white-collar) parents. Also, most of the Politburo members came from humble backgrounds (the above two paragraphs are a summary of chapter 4 of Szymanski 1979).

Trotsky, in spite of the persecution he suffered, continued to believe that the USSR was a workers' state, basically a 'socialist' one, albeit degenerated. His arguments for this position were, firstly, that the means of production remained the property of the state, and, secondly, that the privileges of the ruling bureaucracy could not be passed over automatically from one generation to the next. According to Trotsky, the Soviet ruling bureaucracy was, therefore, not a class but only a caste (cf. Kolakowski 1978: chapters 4 & 5).

Some latter-day Soviet dissidents were of the opinion that, despite everything, there was no new elite class in the USSR. Their main arguments were that, firstly, political power was never handed down to offspring, and, secondly, the state-owned *dachas* and other privileges could not be bequeathed. Zhores Medvedev, a prominent dissident of the early 1970s, said to a foreign journalist: 'A class has to have permanence and stability. Under the old system, the old nobility could be secure in its status. But that is not true now. Everyone is insecure in his position. If he loses it, he loses everything. He cannot pass on his position and his privileges to his children. These are not immutable birthrights' (quoted in Smith 1976: 70).

Another point that should be examined in order to judge whether the USSR was a class society or not is working people's place in society. After all, the power structure of the 'socialist' states was called by 'socialists' themselves dictatorship of the proletariat. Of course, we know that it was a dictatorship of the communist party, not of the proletariat. But was the proletariat in power at least in the production process, in the enterprises? Were the workers themselves determining at least the immediate conditions of their work? Or was there domination and subordination even in this sphere? Here two points are important: the form in which wages were paid and the relations between workers and management.

One form of paying wages had long been disputed, namely the piece-rate. As we know, in capitalist countries, workers hate the piece-rate system, and they hate most those workers who produce above the norm. But in the USSR also, under the dictatorship of the proletariat, the piece-rate system had adverse effects on the morale of the working class, because the official line was that the norms should be frequently revised upwards (Nove 1982: 209).

In April 1918, the Central Council of Trade Unions of Russia issued a regulation in which, in the face of 'economic disintegration' which threatened to bring about 'the extinction of the proletariat', the use of piece-rates 'to raise the productivity of labour' was conceded, albeit grudgingly (Carr 1976: 115). This regulation was strongly criticised by some Bolsheviks known in those days as the Left Opposition as an 'attempt to abolish the eight-hour day' (ibid). But Lenin supported the introduction of piece-rates in the name of 'economic revival and improving labour discipline' (quoted in ibid: 117). Piece-rates became extensive. By 1925, in large-scale industry and mining, they covered 50–60 per cent of workers (Bettelheim 1978: 244). The piece-rate system reduced the actual earnings of a growing number of less successful workers. Many workers resisted these developments. In 1925 – that is, in the NEP period – mass strikes took place without the trade unions or Party organs having been informed. Piece-rates created a split within the working class, in respect not only of earnings, but also of attitudes. At the eighth Komsomol[3] congress (1928), a delegate complained that some workers were 'strutting like peacocks', while others were almost 'beggars' (ibid: 241–51).

During the whole NEP period, the piece-rate system was recognised as a capitalist method and regarded as a transient measure, which had to be accepted for the time being because of the inferiority of the technology (cf. ibid: 244). But in the Stalin era, the system became more extensive. By the late 1930s, almost three-quarters of all workers were being paid according to some variation of this system (Dobb 1993: 464). The view was even expressed that time-rate payments should be abolished. In 1956, 77.5 per cent of all industrial workers were on piece-rate payments (Nove 1982: 209).

After the Stalin era, wage revisions were undertaken, beginning in 1956. Tariff or basic rates started playing a greater role in the earnings of workers. Nevertheless, the percentage of workers on piece rates in 1965 was still 57.6 (ibid: 210). However, easy piece-rates also became one of the instruments used to pay higher actual wages to workers who could not otherwise be recruited or retained.

As regards relations between workers and management, soon after the revolution, in 1918, Lenin said: 'Until workers' control has become a fact ... it will be impossible to pass from the first step ... to the second step towards socialism, i.e. to pass on to workers' regulation of production' (Lenin 1967a: 660). Lenin had also written some time earlier that 'all officials, without exception', should be 'elected and subject to recall at any time' (quoted in Bettelheim 1976: 405).

But, once in power, the Bolsheviks found that these ideals were not realisable. Before the October Revolution, the Bolsheviks supported the factory committees, which were revolutionary in outlook. But soon after the revolution, the factory committees and workers' control created a lot of anarchy in production. Thereafter, taming the factory committees and overcoming workers' control became an immediate goal of the revolutionary regime. This goal was attained by subordinating the factory committees to the trade unions – which were reformist, operated on the basis of industries as a whole and were hence more comprehensive – and later by subordinating the trade unions to the state. Many, especially the anarchists, resisted this policy. One of them characterised the factory committees as 'cells of the coming socialist social order, the order without political power'; another referred to the trade unions as 'living corpses' (quoted in Carr 1976: 111). But they had no strength to prevent anything.

The trade unions were soon reduced to being an organ of the state. It was, of course, declared that among the 'most urgent tasks' of the unions was 'organization of workers' control' (quoted in ibid). But in reality, their chief function was to enforce labour discipline. In March 1918, the seventh congress of the Bolshevik Party demanded 'the most energetic, unsparingly decisive, draconian measures to raise the self-discipline and discipline of workers and peasants' (quoted in ibid: 114). Side by side with compulsion and threats of punishment, which were used during the civil war, there were also moral exhortations and material incentives in the form of bonuses in kind. Many workers themselves showed initiative and enthusiasm to work for 'socialism'.

Two hotly debated issues of the early years were the centralisation of administration and policy making, and its corollary, the one-person management. On these issues, two claims were conflicting: that of efficiency and that of socialist self-government and self-management. Most radical socialists, especially the anarchists, believed that management of the state and industry was well within the competence of any moderately intelligent citizen. But Lenin and many others soon realised that it was not so, that to build 'socialism', 'an immense cadre of scientifically trained specialists' was required (quoted in ibid: 186). In other words, workers were not yet mature enough for self-government and self-management. Hence centralisation of administration in the hands of a bureaucracy was unavoidable.

With regard to the serious problems on the railways, Lenin categorically demanded 'individually responsible executive officials chosen by the railway men's organizations' and 'unquestioning obedience to their orders' (quoted in ibid:191). Generalising the principle of one-person management, Lenin wrote: 'Any large-scale machine industry ... calls for unconditional and strict unity of the will which directs the simultaneous work of ... tens of thousands of people. ... Unqualified submission to a single will is unconditionally necessary for the success of the process of labour organized on the pattern of large-scale machine industry' (quoted in ibid). This policy was bitterly attacked by the Left Opposition, and there was much resistance to it for a certain length of time. Finally, however, Lenin convinced the Party, and the principle of one-person management was accepted.

Henceforth, managers of enterprises were vested with almost dictatorial powers as far as discipline was concerned, and they were not subject to workers' control. Lenin recognised this as an unavoidable 'step backward', but he also considered these measures as temporary and provisional, dictated by specific tasks of the moment (Bettelheim 1976: 156). He thus implied that this strengthening of capitalist relations was to be put right later on. But although by 1926 the difficulties of the earlier years were largely overcome, the same system of management continued. Writing about the slogan of the 'regime of economy', Carr and Davies quote a critical trade unionist who alleged that the managers 'wanted first of all ... to squeeze the workers'. They quote another who complained of the pressure on the worker, who 'feels this pressure on his shoulders', and yet another who complained of the 'pressure on the muscle-power of the workers' (Carr & Davies 1974: 524). In 1928, when the period of planning and forced industrialisation started, one-person management was reinforced. From then on, the main task of the trade unions was to drive the workers to higher productivity (Dobb 1993: 461). In the mid-1930s, complaints became fairly common that unions were not paying sufficient attention to improving the working conditions of their members. This was accompanied by a weakening of internal trade union democracy (ibid: 461–2).

Dobb noted the frequency in 1936 of cases of workers being transferred arbitrarily, of imposition of disciplinary penalties without the accused being given a hearing, of procrastination and a formal–bureaucratic attitude towards suggestions and complaints from workers, and dismissals of workers for 'formal motives' (ibid: 462). By 1937, 'cases were quoted where workers were illegally deprived of their rest-day under the pretext of "voluntary labour" without protest from the trade unions ...' (ibid). Heads of enterprises were always clamouring to intensify labour, raise production norms, reduce wages and strengthen labour discipline (ibid: 219).

After the end of the Stalin era, the situation improved. Basing himself on the research of many Sovietologists, Szymanski wrote that after 1956 the trade unions played a stronger role as defenders of workers' interests, establishing work norms, quotas, standards of labour productivity, ways in which the plan was to be fulfilled, and so on. Workers also had a real say in enterprise decision making, and there was a high rate of participation in general meetings, where the management had to submit themselves to questioning from the floor; such questionings were a real ordeal for the managers (Szymanski 1979: 53–6). But in 1970, 67–68 per cent of unskilled manual workers felt that they had no influence on the affairs of their working collectives. The corresponding figure for managerial personnel and specialists was 13–24 per cent (ibid: 56).

This apparent contradiction in Szymanski's presentation can be explained if one thinks of the concept of alienation. The trade unions could, of course, persuade the workers to come to the general meetings and put critical questions to the managers. But the objective compulsions of a modern, large-scale, industrial economy, which was moreover centrally planned and managed from above, left little room for decision making with workers' participation, except in unimportant matters.

The evidence presented here is not sufficient to assert that the USSR was a class society. At the stage of socialism (before communism), full equality was ruled out even by Marx, who advocated payment according to work done. Although one-person management was not a satisfactory state of affairs for them, the Soviets could not but consider hierarchical forms of work organisation to be more efficient than non-hierarchical and participatory forms. And efficiency was considered by all 'socialists' as well as by all critics of 'socialism' to be very important for the good of society.

The new class and its privileges

There are a few problematic points in the views summarised above. Szymanski stressed the proletarian and similarly humble origin of the majority of the ruling élite of the USSR. But did that guarantee that the USSR was a 'socialist' state and that the working class was in power? Upward social mobility does not necessarily mean that people who have risen to a higher class or stratum remain loyal to and represent the interests of the class or stratum of their origin. A leading Bolshevik theorist knew this. Bukharin wrote in 1922:

> Even proletarian origin, even the most calloused hands ... are no guarantee against turning into a new class. For if we imagine that a section of those who have risen out of the working class becomes detached from the mass of the workers and congeals into a monopoly position in its capacity of ex-workers, they too could become a species of caste, which could also become a '*new class*'. (quoted in Nove 1975: 620)

There were, and still are, 'worker–bureaucrats' in western trade unions. But the danger, according to Bukharin, was greater in the USSR. He wrote:

> The cultural backwardness of the working masses, especially in conditions of general misery, when *nolens volens* the administrative and leadership apparatus has to receive many more consumer goods than the ordinary worker, gives rise to the danger of a very substantial divorce from the masses even of that part of the cadres which emerged from the working masses themselves. ... An appeal to working-class origin and proletarian goodness is not itself an argument against the existence of this danger. (quoted in ibid)

Rakovsky, an exiled Russian revolutionary, wrote in 1929 that the new rulers had 'changed to such an extent that they have ceased, not only objectively but subjectively, not only physically but morally, to be members of the working class' (quoted in ibid).

The next important argument in defence of the USSR was that the members of the ruling élite could not bequeath their power and privileges to their offspring. This was a technical argument, which only forbade one to use the term 'ruling *class*'. But a ruling élite that enjoyed material privileges, and hence was exploiting the working people, was without doubt undermining 'socialism'. The next generation of the ruling élite, although not genetically connected to the preceding one, simply continued the exploitation.

There are two more problems: (1) Szymanski applied the usual sociological categories and criteria in order to find out whether there was an élite class or whether one was forming. But those who maintained that an élite privileged class was ruling the USSR spoke of a *new* class, that is, a class which did not fulfil the *usual* sociological requirements for being called a class. (2) Szymanski and his sources examined the incomes and life-styles of the various social groups or strata on the basis of tariff wages and salaries, and official or officially tolerated scientific publications. But those who maintained that a new exploiting class ruled the USSR spoke of privileges, admitted or unadmitted, and hidden material rewards. It appears to me that the view that there was a strong trend towards egalitarianism in the post-Stalin decades is correct, but only if one examines the incomes of the classes or strata below the new class. The rulers pursued a policy of egalitarianism with regard to workers, engineers, doctors, enterprise managers, and so on. But there was no egalitarianism between the new class and the rest of society.

Those who maintained that there was indeed a new class exploiting the people had a difficulty in proving their point. They had little scientifically generated data. On real income, as distinct from officially recognised income, Nove wrote:

> The proper measure of real income differentials is complicated in the USSR (and nearer home too) by various extras and 'perks': for instance, the use of a car, a better apartment, lavish travel expenses, sometimes the so-called 'envelopes' containing extra banknotes. In addition there is the value of privileged access to goods and services unobtainable by lesser mortals: imported liquors, scarce foodstuffs, good furniture, hotel rooms, the chance to travel abroad. It is hard to know how to take these things into account ... the value of privileged access to goods and services in short supply depends on their being in short supply. ... In the USSR shortages are very much more common, and so special allowance – unquantifiable, unfortunately – must be made for this. (Nove 1982: 215)

Jiri Kosta, who had intimate knowledge of Czechoslovakia, and probably of the whole European 'socialist' block, wrote in 1973:

> That particularly the effective incomes of the highest-level people (leading functionaries) of the Party-, state-, and economic apparatus and some successful artists, sportsmen etc. are kept secret and that they, apart from their high money incomes, are granted material privileges (commodities at reduced price, holiday travels, free services etc.), leads one to the conclusion that the material inequalities have assumed proportions which the privileged groups would like to conceal – obviously out of fear that they would not be able to justify them in front of the masses. (Kosta *et al.* 1973: 175–6)

Rudolf Bahro, who had intimate knowledge of the GDR, wrote in 1977:

> The differentiation of income among the mass of those involved in the reproduction process in its present form ranges in the GDR between say 500 and 1500 marks. Anything above this goes to people who are already privileged in other ways, and thus simply receive an additional reward for their bureaucratic obedience or for special services to the ruling power. Salaries or incomes of 3000 marks and upward can only be based on the exploitation of other people's labour, given our present average income. (Bahro 1981: 386)

Milovan Djilas wrote of the country homes, furniture, best housing, cars and similar things. He also wrote of the privileges and gifts from the state – such as generous housing – which were 'the everlasting and inexhaustible sources of benefits to the political bureaucracy', a stratum to which people like the local Party secretary and the chief of the secret police belonged (Djilas 1958: 57). He gave a concrete figure, quoting one Orlov, who wrote in his book *Stalin au Pouvoir* (Stalin In Power), published in Paris in 1951, that in 1935 the average annual pay of a Soviet worker was 1,800 roubles, whereas pay and allowances of the secretary of a district committee of the Party amounted to 45,000 roubles a year, a ratio of 1:25 (ibid: 46).

Szymanski wrote that it was no secret that some people in the USSR had an income ten times the average industrial wage – prominent artists, writers and performers, leading university administrators, professors and scientists (Szymanski 1979: 63–5). But Djilas did not include such people in the new class. Moreover, the masses probably did not mind if an author earned a lot because many millions of copies of his or her books were sold, or if a famous musician became rich because he or she was often invited to perform in the Western capitalist countries. Their objections were to the effective incomes of the high-level functionaries of the Party, state, and economic apparatus.

Djilas meant by the new class the 'bureaucracy, or more accurately, the political bureaucracy' (Djilas 1958: 38). He explained: 'However, ... only a special stratum of bureaucrats, those who are not administrative officials, make up the core of the governing bureaucracy, or, in my terminology, of the new class. This is actually a party or political bureaucracy' (ibid: 43). In his understanding, 'The new class may be said to be made up of those who have special privileges and economic preference because of the administrative monopoly they hold' (ibid: 39). The newness of the new class consisted, according to Djilas, in the fact that its members did not formally own the means of production. But informally, the means of production were the collective property of the new class. It could use this property, derive benefits from it and dispose of it in the way it liked. The new class was, according to Djilas, also an exploiter class. And he believed that it was parasitic. That is logical, because this class, according to him, was not performing any regular (administrative) work. Djilas wrote that the new class was 'voracious and insatiable'. It knew that the privileges came only through power. Consequently, 'unscrupulous ambition, duplicity, toadyism, and jealousy inevitably must increase. Careerism and an ever expanding bureaucracy are the incurable diseases of Communism ... unscrupulous ambition must become one of the main ways of life ...' (ibid: 60–61).

In the 1970s, during *détente*, when many foreigners lived in Moscow, and Soviet citizens were not too afraid to talk to them, much more information on the privileges of the new class was available. Hedrick Smith, correspondent of the *New York Times* in Moscow in the first half of the 1970s, gave many details, which cannot be presented here for lack of space (see Smith 1976: chapter 1). Apart from the unjustifiable material privileges, the behaviour of members of the new class in their relations with lower-placed citizens resembled the typical attitudes of an aristocratic ruling class in a feudal society. The arrogance of VIPs and their chauffeurs was notorious. Smith told the story of an army officer in Tashkhent, who simply ignored the people waiting in a long queue for a taxi, placed himself at the front and took the next one. It happened very frequently that people who had reserved seats for a flight were told at the airport that they would have to take the next plane because some Party bosses had taken their reservations. Smith told the story of Ludmila, Kosygin's daughter, who, while travelling with her family by train, made the railway personnel serve her food in her cabin, although the rule was that passengers had to go to the dining car. He told the story of Galiya, Brezhnev's daughter, who bought from a shop a piece of art which had earlier been declared not for sale (see ibid). Komarov told stories of army generals and other highly placed people who violated all rules and orders, and held hunting parties in nature reserves, where hunting was prohibited, such as on the shores of the Caspian Sea, or killed protected animals such as polar bears (Komarov 1980: chapter 6).

Reactions of the masses

How did the masses react to these privileges? One might say that, compared to Western capitalist societies, the chasm in the Soviet Union between the highest and the lowest strata was not too great. That might have been true. But in every society, people judge such things in terms of the values of that society. In the USSR and other 'socialist' societies, at least the cherished values were socialist. Otherwise the new class would not have tried to consume its luxuries in secret.

It was risky to criticise the leaders. Nevertheless, many people did comment on and criticise the behaviour of the new class, sometimes to foreigners. Smith quotes many such comments (the following details are from Smith 1976: chapter 1). The most telling among them was the following joke: Brezhnev wanted to show his old mother how much he had achieved in life. He showed her his luxurious apartment in Moscow. She did not say anything. He then showed her his Zil limousine. She did not say anything. He then drove her to his *dacha*. She did not say anything. He ordered a helicopter and flew her to his country house and showed her all its splendours: the big dining hall, the beautiful fireplace, his collection of guns. But she did not say anything. Brezhnev became very impatient. He wanted to hear some praise from his mother, so he finally asked her: 'Tell me mama, how do you find all this?' His mother replied with a little hesitation: 'Well Leonid, everything is fine. But what will you do when the Reds come back?'

Smith reported:

> Russians themselves comment that the upper class feeling today seems increasingly like Russia before the Revolution. An engineer observed to me that what Marx had predicted for capitalist society – increasing concentration of economic power in fewer and fewer hands and a widening gap between the elite and the masses – seemed to be happening in the Soviet Union today. (ibid: 69)

Smith reported that many Soviet citizens were enraged at the existence of shops in which only foreign currency was accepted. A Russian said angrily: 'It is so humiliating, so insulting to have stores in our country where our own money is not valid' (ibid: 44). Between the young men of Zhukovka village, near Moscow, and the sons of the new class, many of whom lived with their parents in the VIP *dachas* nearby, there were often fights and quarrels that assumed the character of class struggle.

He had the impression that most Soviet citizens hated the privileges of the élite, found their standard of living unseemly, a mockery of 'socialist' ideas. This disturbed intellectuals particularly. Smith told the story of a poet's wife who was a guest at a party given by Dmitry Polyansky, the Minister of Agriculture and Politburo member. She went to the bathroom and alarmed fellow guests by the noise she made as she proceeded to smash bottles containing French perfumes, exclaiming: 'The hypocrisy of it all! This is supposed to be a workers' state, everybody equal, and look at this French perfume!' (ibid: 73). But, according to Smith, most people felt helpless and became cynical or fatalistic. They did not try to change the situation. They wanted only to belong to the privileged and have a piece of the cake for themselves.

MORAL DEGENERATION

The formation of a new class with power and privileges was the strongest proof of moral degeneration in 'socialist' Soviet society. But there was also moral degeneration in a narrower sense among the whole people, including workers and rank-and-file communists. It is probable, however, that the formation of a new privileged class to a large extent caused this wider moral degeneration.

The idealism of the early years

The early years of the regime were marked by revolutionary idealism. In 1919, the revolutionary ardour of several hundred workers inspired them to volunteer to put in an extra six hours after work on Saturday. This practice, called 'Communist Saturday', spread; it was a form of voluntary unpaid mass labour undertaken for the rapid completion of important tasks, including repairs. There was another institution called 'shock work'. In 1920, especially active and highly motivated workers were assigned to especially difficult or urgent tasks. In 1926, workers, especially young workers and their organisation, the Komsomol, initiated the Socialist Emulation Movement in order to speed up the growth of industrial production. It took the forms of Communist Saturdays, and undertakings to increase production or to carry out exceptionally difficult and urgent tasks, these undertakings being made by teams of workers called shock brigades. Another form of workers' initiative was 'counterplanning'. It started in 1930. Workers would criticise the plan and make counter-proposals with much higher targets, pledging at the same time to overfulfil the plan. In 1935, enthusiastic young workers initiated the 'Stakhanov movement' to rationalise working methods, which resulted in a tremendous increase in labour productivity.

In her book on the Stalin era, Anna Louise Strong, who visited the USSR several times and also lived there for some time, wrote the following on 'new people':

> The characteristic of the people who built the new industries and farms was boundless initiative. ... never in any land, until my later visits to China, have I met so many dynamic individuals as those who found expression in the USSR's five-year plans. ...
>
> This zeal to create filled not only the leaders. It was born in millions of plain citizens as they saw new roads to life. ... once illiterate peasants became farm scientists, amateur actors, parachutists, aviators. (Strong 1956: 46)

Even foreigners who were helping were inspired by this spirit. Strong quoted an American who said: 'The greatest thing in life is work. No not just work. Creation! And in this spot of time in which we live, there is the chance to create without end or limit. Could I turn from an hour of creation to be nice to a wife or to come to dinner on time?' (ibid).

But all this enthusiasm and idealism gradually faded away. Among the reasons were 'the development of administrative centralism, multiplication of rules and constraints imposed by the state (not propitious for initiatives from below), and penetration of bourgeois specialists into the state machine, with the resulting bureaucratisation' (Bettelheim 1976: 201). In 1927, the title 'Hero of Labour' was created, which was no mere title. Attached to it were several material advantages. Later, the special bonuses, the higher earnings (due to the system of piece-rate and progressive premiums) and the honours the Stakhanovites got, placed them at a much higher economic and social level than the majority of the workers. 'Gradually, emulation came to have the effect of setting against each other different groups of workers, and even individual workers; the best performances were used by heads of enterprises to revise work norms upwards and increase the intensity of labour' (Bettelheim 1978: 253). Among workers, discontent, indifference, even hostility set in. 'Enthusiasm did not entirely evaporate, of course, but it became confined to a minority ...' (ibid: 255). No wonder managers were complaining about labour indiscipline.

When, in the early 1920s, the Communist Saturdays gradually faded away, there were objective reasons. In the NEP period, when capitalist relations of production were revived, the objective situation was not favourable to the growth of idealism. But Lenin had recognised in the Communist Saturdays 'the cells of the new, socialist society'. In his pamphlet *A Great Beginning* he wrote:

> It is the beginning of a revolution that is more difficult, more tangible, more radical and more decisive than the overthrow of the bourgeoisie, for it is a victory over our own conservatism, indiscipline, petty bourgeois egoism, a victory over the habits left as a heritage to the worker and peasant by accursed capitalism. Only when this victory is consolidated will the new social discipline, socialist discipline, be created; then and only then will a reversion to capitalism become impossible, will communism become really invincible. (Lenin 1967b: 223)

The reality in advanced 'socialism'

From the idealism and enthusiasm of workers for building 'socialism' in the early years of the revolution, let us now come to the state of morale and morality a little over one decade before its collapse. Let us begin with the people in charge of the economy and the working collectives in the enterprises.

Standard critics of 'socialism' – both pro-capitalist ones and 'socialist' reformers – have seen the roots of the undisputed inefficiencies and weaknesses of the 'socialist' economies in the very system. By 'system' they meant both the basic system – state or collective ownership of the means of production, co-operation instead of competition, and planning – and the *particular* systems of centralised planning and management, and of incentives. Nove wrote:

> these distortions are usually unintentional. The planners desire that footwear should fit to customers' tastes and feet. They most certainly desire that rolling-mill equipment be modern and of high quality. If the opposite occurs, this is ... due ... to a malfunctioning of the system. ... this is a systemic malfunctioning, not due to errors of individual planning officials or managers. (Nove 1982: 119–20)

Elsewhere Nove wrote: '... and there is about the whole system an unintended discouragement of both innovation and quality ...' (ibid: 175). Conert concluded that the problems were due to 'defects *of* rather than defects *in* the system of economic administration in the USSR' (Conert 1990: 27).

The critics have given innumerable examples of inefficiencies, weaknesses, wastage of material and labour resources, etc. (Unless otherwise stated, the following description of the problems of the Soviet economy is based on Nove 1982, Conert 1990, and Aganbegyan 1988). But I am not convinced. I think their analysis is wrong, because they mostly confuse the effects with the causes.

All critics have elaborately described the difficulties of the planning and production processes, which can be summarised as follows: detailed centralised planning is an impossible task because it is too big. It requires an enormous quantity of detailed information, which the planners could neither collect themselves nor get reliably from below. To check whether the enterprises were working according to plan, the planners, the economic ministries and other government offices needed enormous quantities of information from both enterprises and inspectors. This paperwork claimed much of the precious time and energy of managers, engineers, accountants, and so on, which resulted in less attention being devoted to the actual production process. It was also useless: the greater part of such reports could not be read and very little of the data was actually processed. Centralised planning and management also meant that too many interventions were made from above in the day-to-day production process. Plans were changed too often, orders came from above that contradicted previous orders, and so on and so forth.

Because of the enormity of the task, plans, especially the operational yearly plans, generally reached the enterprises after the beginning of the plan period, and were not sufficiently balanced. Enterprises, therefore, often did not know exactly how much they should produce. As a result, they had difficulty getting the inputs in time, and if they did not get the right inputs, the quality of their products suffered. Quality suffered also because planners laid the greatest stress on achieving quantitative growth and because it is anyway almost impossible to plan quality and the exact product mix of an enterprise that produces several things. The 'socialist' economies were, on the one hand, known as 'economies of shortages'. On the other hand, they produced many unnecessary things because of faulty planning. Thus, for example, too many tractors were produced and thrust upon the collective and state farms.

Critics generally said that a socialist economic system must unavoidably suffer from lack of motivation to work hard and well. Since in a planned economy everything produced is also sold, since the possibilities of making high profits and going bankrupt are absent, since the state guarantees that nobody is unemployed, and since income differentials are low, nobody can have much motivation to work hard. That was, according to the critics, also the cause of innovation weakness in the 'socialist' economies. Enterprise managers were neither motivated to make their own innovations nor eager to introduce those made by others.

At first sight, this analysis appears correct, especially when many examples are given in its support. But a closer look at the examples reveals some deeper causes of inefficiency and weakness. Socialists, and 'socialists', wanted to build an ideal socialist society, not just an efficient economy. This meant that the policy makers and planners wanted rapidly to eliminate the difference in the standard of living between town and country, and between developed and underdeveloped regions. In the USSR, it meant that they had to devote disproportionately high percentages of investment to the development of, for example, the backward central Asian republics. This meant in many cases that they had to deliberately contravene some of the objective laws of economics. For instance, they had to forgo the economic advantages of concentration. Writing in 1980, Nove testified that resources had actually been redistributed towards the more backward areas, that wage rates in central Asia were similar to those in central Russia, that prices of central Asian products were relatively favourable, and that social services there had been on the standard Soviet scale (Nove 1982: 69).

Nove points to the negative effects of this policy on economic efficiency by speaking of 'a contradiction between economic advantage and the "regional" objective' (ibid: 70).

> the central Asian republics lack fuel (except gas), they are remote from the principal consuming centres, transport costs are high. Economies of scale and complementarities are more readily to be found in or near existing areas of industrial concentration. Planners must also take the location of natural resources into account, and these point (for instance) to Siberia rather than to Tadzhikistan as priority areas for large-scale investments. (ibid.)

Given their ideal, and given the objective material circumstances described above, it was impossible for the Soviets either to leave the allocation of scarce goods to the forces of supply and demand in a free market or to plan it according to the narrow principles of purely economic advantage. It had to be planned politically and by directives. Otherwise more developed nationalities and regions, and strong enterprises, would have got hold of most of the scarce resources and economic benefits. The leaders had to sacrifice optimal allocation and efficiency to some extent for the sake of distributive justice. This was rationing, resorted to even in capitalist countries in times of crisis.

The positive discrimination in favour of the backward republics and the fact that in the post-Stalin decades income differentials among the great majority of the people were reduced, show that, at least at the official policy levels, the Soviet 'socialist' society was trying to live up to its ideal. But many facts from the economic sphere reveal a different picture; simultaneously, they reveal the deeper cause of the inefficiencies and weaknesses of the economy.

Uncomradely communists

The standard critics were, of course, right to say that the information the planners received from the enterprises and economic ministries was unreliable. But why was it so? The answer is that the latter deliberately supplied misinformation to the former. They concealed their real production capacity and demanded quantities of inputs higher than what was needed. They hoarded inputs, including labour capacity. The purpose of these manipulations and dishonest activities was to be able to fulfil and overfulfil plans easily. This was very important for the working collectives (from the enterprise director to the lowest-ranking worker) because a substantial part of their income consisted of bonus, paid only if the plan was fulfilled or overfulfilled. No planned economy can function efficiently if its database is wrong.

The planners knew about the manipulations. They made cuts in the input demands and, as regards plan targets, they followed the 'ratchet principle', taking the previous year's performance as the base and raising it by a certain percentage for the next year's target. But this made fulfilling plans more difficult for the enterprises, and the working collectives feared that they would get no bonus. This was the reason that a working collective, even when it overfulfilled its plan, never overfulfilled it to the extent it could have. It was a vicious circle, the two sides distrusting each other. Most of the people involved in it were members of the Communist Party, supposed to be comrades, but they did not trust each other.

The result of this 'unsocialist' mental attitude was very bad. There was a well-known cartoon showing a single huge nail in a factory; the manager was pointing at it and saying, 'the month's plan fulfilled'. The plan was fulfilled in tons. Indeed, Soviet sheet steel was notoriously thick; sheet glass was too heavy and paper too thick when they were planned in tons. Even chandeliers, planned in tons, were unnecessarily heavy. When a production target for cloth was stated in length (obviously a mistake), it gradually became too narrow. When later it was stated in square metres, the quality deteriorated. If production targets were stated in numbers of machines, there was a shortage of spare parts. If, in the case of transport enterprises, plan targets were set in ton-kilometres, optimal and cheap transport possibilities were neglected.

When the monetary indicator 'gross output' was used, the value of unfinished products was included. And because it included the value of inputs, managers demanded costlier inputs than necessary and tried to get the prices of their products fixed on the basis of those of the inputs. Here are two examples: catering enterprises preferred to fulfil their plans in monetary terms by selling bottles of vodka rather than by serving tea, which was much cheaper and required more labour; construction enterprises improved their labour productivity by using costlier materials, because labour productivity was calculated as money-value of materials used per unit of labour. Referring to such phenomena, Gorbachev wrote in 1987:

> The worker or the enterprise that had expended the greatest amount of labour, material and money was considered the best. ... It became typical of many of our economic executives to think not of how to build up the national asset, but of how to put more material, labour and working time into an item to sell it at a higher price. (Gorbachev 1987: 5)

Tatjana Saslawskaja knew the explanation. With reference to the price determination process of new products, she wrote in 1989: '... there arises an interest in the enterprises to estimate and maintain the production costs as high as possible, because the profit margin is calculated as a percentage of the production costs' (Saslawskaja 1989: 100). An enterprise's profit was one of the important factors that determined the income (through the bonus) and the career of its managers.

This attitude was also partly responsible for the innovation weakness of the 'socialist' economies. Very often, one enterprise did not want to take up an innovation from another. For if it did, its actual productivity and/or production capacity, which it always tried to conceal, would have become known to the planners, at least to the extent that they were connected with the use of machines containing the innovations (Lohmann 1985: 74–5).

Lack of co-operation between communists was also manifested in the phenomenon of the sellers' market. To a large extent, the exact product mix of enterprises was to be the result of negotiations between suppliers and takers, the takers saying what they needed and the suppliers saying what they could supply. In many cases, the suppliers adopted a take-it-or-leave-it attitude, because they were in a stronger position in an economy of shortages. That explains why a factory that got, say, the planned quantity of rolled steel did not get it in the required profile.

Egoistic, 'unsocialist' behaviour and ignoring the national interest were visible even at higher economic levels. Since getting the required inputs was a problem, economic ministries ordered 'their' supplying enterprises to prioritise the requirements of 'their' buying enterprises, often leading to supplies to enterprises belonging to other ministries being delayed or not made. In fact, the economic ministries were notorious for their particularistic, empire-building attitudes. They tried to get as high a portion of the national budget as possible. They preferred to submit proposals in those priority areas of investment in which proposals could gain acceptance by the planners even before the completion of the cost–benefit analysis. After acceptance, the ministries hurried to start the large-scale construction work. If the cost–benefit analysis subsequently raised doubts about the rationality of the project, the ministries pointed to the large sums already spent and argued that the project should be completed.

The situation did not improve when Khrushchev abolished the economic ministries and transferred their functions to regional councils. The latter also indulged in empire-building, ignored the national interest, diverted resources to local and regional needs, and neglected their supply obligations to enterprises in other regions. There was also distrust and lack of co-operation between the economic ministries and their own enterprises. The former tried to lure the latter, with the prospect of extra bonuses, to submit counterplans with targets higher than those they had been given. But the ministries then took every such counterplan to be proof of successful concealment of capacity. So they often revised the obligatory plan upwards by adding half of the difference between the original target and the counterplan target, to arrive at a higher target for the obligatory plan.

To sum up, the bonus fixation of the working collectives and the particular interests of the economic ministries resulted in a vicious circle of distrust, cheating, and lack of co-operation. Such things are also present in all capitalist societies, and to a higher degree. But one expected a different behaviour pattern from communist economic executives in a 'socialist' society. When this expectation was betrayed, there was no chance for 'socialism', which was actually a moral project, to succeed.

One question that now arises is why, in the post-Stalin decades, nobody was punished for the dishonesties described above. The answer seems to be, firstly, that such behaviour had become almost normal and, secondly, plain cowardice. Everybody was locked into a corrupt interdependence. A network of informal relations had developed, the motto of which was: you protect me, then I protect you. A second question that arises is what the Communist Party and its leading organs and members were doing. The party organised many kinds of campaign. But we know that they did not work except under some kind of compulsion. Peter Strotmann commented in 1969: '... in a hierarchically built system of privileges and bureaucratic institutions, ... every campaign that tries to generate activities of the masses without the promise of material advantages, must appear to be hypocrisy and must also be recognized by the masses as such' (Strotmann 1969: 24).

'Socialist' reformers have raised two questions: would it not have been better, even in a planned 'socialist' economy, to give the management at enterprise level more freedom and decision-making power? And would it not have been better to make more use of market mechanisms without wholly giving up 'socialist' planning? I shall discuss these questions, which are more technical in nature, in chapter 6. I do not think that the failure of 'socialism' was caused to any significant extent by negative answers given to them. The negative answers might have increased the inefficiency of a basically unsound system pursuing impossible goals. But positive answers would not have been able to prevent its ultimate failure. What I have written above and what I shall write below in this chapter on morale and morality in the USSR shows that there the very spirit of 'socialism' was dead.

Corruption in everyday life

Smith gives a vivid description of the situation in a chapter on corruption, based to a large extent on Soviet press reports, and on conversations with his Russian friends and acquaintances (unless otherwise stated, the following is a summary of Smith 1976: chapter 3). Towards the end, in the 1970s, it was indeed very bad. Smith wrote: '... Russians insisted that it [corruption] had risen sharply as Soviet society became more affluent in the late Sixties and Seventies' (Smith 1976: 115). He gives innumerable examples of theft of state property. Chauffeurs of state-owned cars stole petrol; drivers of lorries with cement-mixers stole cement; sales personnel in shops stole meat and fruit and so on; workers stole spare parts for cars and other consumer durables. For all stolen things, which were scarce, they found consumers eager to buy them at a much higher price than in a shop. Or they presented them to people from whom they expected something in return. In September 1972, *Pravda* revealed that in the Russian Soviet Republic, in a relatively short period, more than 200 large-scale thefts of state property were uncovered and that more than 50 per cent of these cases were each part of long-term operations of well-organised criminal syndicates. On 1 January 1975, *Izvestia* reported that in 1972–73 more than a third of Soviet car owners used petrol stolen from the state. Others calculated on the basis of Soviet statistics that every year 560 million litres of petrol were being stolen. The thieves and the purchasers of stolen goods were not only people from the lower strata. Among them were a chemist in a research institute, an engineer in a car factory, and a dentist.

Apart from theft, there was also widespread corruption: students, who got subsidised train and air tickets, sold them at a higher price. The director of a big supermarket sold scarce but highly coveted food items to her friends, who returned such favours by presenting her with, say, two tickets for a sold-out ice-hockey match. She also took cash bribes. In Baku, the police demanded a bribe of 500 roubles to grant a building site for a private house, and a job as a taxi driver cost 400 roubles. A surgeon in a state hospital had to be bribed 50 roubles to get an operation done, and a good dentist 150 roubles for good crowns.

Theft, all kinds of corruption, and illegal economic activities together formed a counter-economy, which was ironically called by Russians 'creeping capitalism'. It included an additional occupation, whereby the main job was neglected or badly done. It also included illegal enterprises producing scarce goods and services for the black market. Some of these enterprises were within regular state-owned factories. Andrei Sakharov's estimate of the extent of the counter-economy was at least ten per cent of GNP. Others gave both lower and higher figures (cf. ibid: 114) In fact, the formal Soviet economy was to a considerable extent dependent on the informal. The press often carried reports of collective farms and other enterprises that needed (stolen) materials and illegal services from the counter-economy in order to be able to fulfil their plans. The efforts of the state to check economic crimes were not very successful, because police officers and economic inspectors often worked for the criminals. By the 1960s, there was already a mafia in the USSR.

After reading all this, one cannot but come to the conclusion that probably the majority of the Soviet people had lost all sense of dignity, to the extent that individuals such as the chemist in a research institute could talk to a foreigner about their own corrupt activities without shame. One can argue that much of the counter-economy would not have been illegal and criminal if the state had allowed more private enterprise. But then such a great part of the economy would have been capitalist that the USSR would not have remained a 'socialist' state any more. The Bolsheviks had not made the October Revolution merely to build a mixed economy.

The Party and its members

As far as the communists themselves were concerned, in 1919 a decree provided that 'responsible political workers' like the People's Commissars were to be paid 2,000 roubles per month, which meant partially abandoning *partmax* (Bettelheim 1976: 165). This was more than three times the minimum wage of a worker. This example obviously set a precedent. In 1926, technicians and managers who were Party members were getting on average more than 14 times the minimum wage of a worker (Bettelheim 1978: 249). Whatever little of *partmax* was left was abolished by Stalin after 1931 (Nove 1972: 208). This meant that higher-level Party members were actually cut off from the working class, even if they had come from it. They often tended to form cliques whose members covered up for each other – what were called in the USSR 'family circles' (Bettelheim 1978: 336).

The withering of inner-party democracy and the centralist style of party administration 'undermined the quality of relations between the different levels in the Party, between the rank-and-file and the top leadership and between the political and administrative leadership' (Bettelheim 1976: 425). Relations between Party members generally became more abrasive. But it was Stalin's reign of terror that inflicted the greatest damage on the morale and morality of the communists and on the image of socialism or communism. The fate of the known leaders of the Party has been recorded carefully. We know, for example, that out of the 26 members of the Central Committee elected at the

11th Party Congress (1922) twelve were executed, two disappeared, two committed suicide, one was sentenced to 20 years imprisonment (all between 1936 and 1938), and one – Trotsky – was murdered in 1940. Of the 17 members of the Politburo elected at the 15th Congress (1927), nine met with similar fates (Ellenstein 1977: 114–15).

The point is that this great crime finally and completely destroyed the spirit and morale of the Party. All honest, upright, sincere and courageous communists were executed, imprisoned or deported. This conclusion is logically valid, because anybody who did not protest against these atrocities was not upright, and anybody who protested was executed. Those who were left were accomplices, cowards, and careerists, who threw themselves at the feet of the great dictator. The vast majority of both the Party members and the masses were stunned, and learned to keep quiet, conform, and save their own skins by denouncing others. A whole generation was thus lost. Nobody was left to protest, take initiative, or criticise. What remained was a monolithic block of obedient servants, lackeys, and lickspittles.

The victory of the USSR over the Nazis proves only that the Soviet people, demoralised after having failed to build a socialist society, found in the Second World War another great cause to live and die for – the defence of Mother Russia. Nobody ever denied the greatness of the victory. But nationalists in other countries and at other times have fought as heroically. In fact, as far as the Red Army was concerned, foreign observers had noted by 1928 that both communist and non-communist army leaders were united in an uninhibited faith in Russian nationalism (Carr 1971: 320). After the war, in the days of the cold war, a 'sick nationalism' developed, 'an irritated, excessive patriotism, which denounced as "cosmopolitanism" – and almost as treason – any belief that any land but Russia had ever invented anything good' (Strong 1956: 109).

Amalrik wrote in 1969:

> The need for an ideological underpinning forces the regime to look towards a new ideology, namely Great Russian nationalism. ... The need for a viable nationalist ideology is not only acutely felt by the regime, but nationalist feelings also appear to be taking hold in Soviet society, primarily in official literary and artistic circles. (Amalrik 1980: 40)

No genuine socialist or communist needs to be told that nationalism is the negation of socialism/communism, which is unthinkable without internationalism. 'In reality', wrote Djilas, 'national Communism is Communism in decline' (Djilas 1958: 190).

Smith reported on the morale and morality in the Party in the first half of the 1970s (unless otherwise stated, the following material is from Smith 1976: chapters 3 & 11). He observed the open cynicism among the Party members with whom he could talk and about whom he heard from his contacts. Two of his acquaintances admitted that they had become Party members because they wanted to get on in life and to travel abroad. A Party functionary told Smith: 'Nowadays, people become members practically without any reason. ... It is like going to the Church for people in the West: one does it out of habit, not out of conviction.' According to a teacher who had to attend political meetings, not even those communists who gave lectures at such meetings had any faith in their ideology. A young Party functionary said: 'Nobody believes in the ideology any more. Nobody needs it any more.' Asked why the Party had any interest in him at all, he said: 'The Party knows that it has you in its grip once you have joined. One who is not a member can refuse to do certain things that the Party demands. But once you have become a member, you must do what the Party says. Discipline, you know.'

Blind loyalty to one's leader or head of department was the rule in Soviet society. For one's career, it was important to understand, or better still to anticipate, the wish of the leader or the chief. People with original ideas, initiative, or superior knowledge were disliked. A leader or chief also had the power to grant his/her followers or subordinates foreign travel or a larger apartment, or to select one for a higher post, even if that person was useless.

Smith also gave many examples of widespread corruption among communists – not only among rank-and-file or middle-level cadres, but also among top-ranking communists. An engineer, a member of the Party, said: 'They do it so openly and so shamelessly. In our city, the Party bosses need only give a ring to the fur factory, and they get the best fur coats sent to their homes free of cost.' The Party functionary who was in charge of the Party apparatus' special shops in Moscow had become a millionaire through his underhand dealings. Komarov wrote that large-scale poaching in nature reserves was possible because high Party and government officials took regular bribes (Komarov 1980: 86).

As a principle, investigations and cases involving leading Party members were kept secret in order to protect the image of the Party. But, as a high functionary said in a very worried tone, 'How can the Church survive if the Pope and the cardinals are venal Philistines?' This state of morale and morality in the Party naturally had its corresponding effect on the whole people.

One might now suppose that the people wanted to get rid of the system. But Smith reported that in spite of all these facts, the broad masses, unlike the intellectuals, still believed in the system. A foreman in a factory told Smith: 'The workers revile and sulk, but they revile individuals. I have never heard that anybody had attacked the Party or the System.' Smith also quoted an interpreter who told him that the people did not even look at the huge banners the Party used to put up during campaigns; but he said later: 'Our ideal, the ideal of socialism, the ideal of the people working for the general weal, is indeed far superior to your profit-orientation, although we ourselves see that this ideal has not been realised up to now.'

CRISES AND TRANSFORMATION OF SOCIAL FORMATIONS

Elmar Altvater (1992, 1993), in explaining the failure and collapse of 'socialism', differentiates between small and big crises. In capitalism, small crises function as negative feedback that makes self-correction possible. They are banal, they are resolved within the given social, political and economic structures, their institutions, and their forms of operation, which are not questioned. But a series of small crises can become uncontrollable within the given institutions. When that happens, protracted and profound social, political, and economic transformations begin. Institutions themselves become affected. This process signifies the end of an era. Such crises can be called a crisis of the social formation. Capitalism can overcome even these crises: thus it could survive the great economic crisis of 1929–33, fascism, and the Second World War. It could transform itself into the present social formation known as 'Fordism'.

Now the question arises as to why the social formation 'socialism' could not overcome its own big crisis in the 1980s. Altvater's answer is that in 'socialism' there were no small crises. There were many deficiencies, but they did not or could not come to the surface, not in the radical and effective form of a small crisis as in capitalism.

> In the languid flow of crisis-free development in 'actually existing socialist' societies, which are nevertheless full of contradictions, so much destructive potential sinks to the bottom as sediment that the river comes to a standstill. The social stagnation becomes a social explosive force and, in the course of a big crisis, pushes the destructuring much beyond intrasystemic transformations. Then what takes place is not … transformation, but a revolution. What comes after that is uncertain. Revolutions are not *per se* progressive; in history there is also regressive involution. (Altvater 1992: 63)

When, before the collapse of the system, a legitimation crisis arose, there was no mechanism for transformation. The social formation was too rigid. It had lost the capacity to evolve.

Many people in the 'socialist' societies had noticed the signs of the coming big crisis. Reforms had been tried. But they all came from above. They were bound to fail, because both adaptation and transformation can take place, according to Altvater, only through self-organisation of the concerned society. But self-organisation in the 'socialist' societies was not possible. Individuals were demoralised and frustrated, bothered only about their private life: they had become lethargic in social matters, and social institutions had lost their cohesive force.

In the second half of the 1980s, such a transformation was tried in the USSR through *perestroika* and *glasnost*. The masses made the best use of the opportunity. When they, as voters, got a chance to choose between two or more candidates, they voted in large numbers against those of the Party, many of whom were defeated. At the same time, they resisted efforts to dilute the 'socialist' character of their economy. They did not realise that in view of the limits to growth, their kind of 'socialist' economy had become impossible. Altvater thinks the explanation of capitalism's ability to transform itself is the combination of market economy and democracy. But can this combination help humanity to overcome the great ecological crisis of today and the consequences of the limits to growth? Which transformation must the human societies of today undergo: towards eco-capitalism or towards eco-socialism? I shall discuss these questions in the following chapters.

THE TRAGEDY OF THE COMMONS

I have devoted two chapters to the search for an answer to the question why 'socialism' failed. Each deals with one part of the search and the answer. But the whole can be summed up in the concept of 'the tragedy of the commons', first used by Garrett Hardin in 1973. The idea is very old. Aristotle said: 'What is common to the greatest number gets the least amount of care' (quoted in Hardin 1980: 115). But Hardin used the idea in a way highly relevant to the problems we are discussing. We can conceptualise the failure of 'socialism' as degradation of a commons.

There are substantial parallels between the abstract pasture and herdsmen in Hardin's essay and the concrete resources and real-life economic actors in the former USSR. The Bolsheviks transformed the old Russia into a huge commons in order to create a 'socialist' society. All resources belonged to the state and were used by the economic actors without having to pay a price. The USSR became a commons in both of Hardin's meanings: as a source of raw materials, land, and water (Hardin used the term 'food basket'); and as a sink for all kinds of pollutants ('cesspool'). The people – every enterprise, every government agency or department, and every individual – treated the country's resources as Hardin's abstract herdsmen treated their abstract pasture. Not only the natural resources, but also the state with its funds, equipment and manpower, enterprises with their machines and materials, and all infrastructures and utilities were viewed as commons. And they were all plundered and misused for unsustainable and illegitimate ends.

Hardin wrote: 'It is to be expected that each herdsman will try to keep as many cattle as possible on the commons' (Hardin 1973: 137). Like the herdsman, the top political and economic leaders of the USSR tried to produce as much as possible, and the whole population wanted to consume as much as possible. Hardin wrote:

> As a rational being, each herdsman seeks to maximise his gain. ... he asks, 'what is the *utility to me* of adding one more animal to my herd?' This utility has one negative and one positive component. ...
>
> Since the herdsman receives all the proceeds from the sale of the additional animal, the positive utility is nearly +1.
>
> Since, however, the effects of overgrazing are shared by all herdsmen, the negative utility for any particular decision-making herdsman is only a fraction of –1.
>
> Adding together the component partial utilities, the rational herdsman concludes that the only sensible course for him to pursue is to add another animal to his herd. And another; and another ...' (ibid: 137–8)

Unlike Hardin's herdsman, the managers and workers of an enterprise in the USSR did not get all the profits from their economic activities, but a portion of it in the form of bonus. This portion was so substantial that almost all their actions and reactions were strongly influenced by the question: how can we maximise *our* bonus?

Bonus Maximization

On the ultimate result of the herdsmen's behaviour, Hardin wrote: 'Such an arrangement may work reasonably satisfactorily for centuries ... Finally, however, comes the day of reckoning, that is, the day when the long-desired goal of social stability becomes a reality. At this point, the inherent logic of the commons remorselessly generates the tragedy' (ibid). For the USSR, the day of reckoning came at a time when the society appeared to have achieved a substantial degree of social stability. In the second half of the Brezhnev era, in the 1970s, there was *détente* in its relations with its former enemies, even economic co-operation. The USSR achieved enormous power and prestige in the international arena. The standard of living of the people had risen substantially.

Dissidents were only mildly punished; nobody was executed. The vast majority of the people accepted the system. But this was also roughly the time when the economic decline began, manifested in declining growth rates and a rapid worsening of the state of the environment. Nevertheless, the vast majority of the people and all their leaders, like Hardin's herdsmen, wanted more, and a higher standard of living, to catch up with the USA, until the collapse came less than a decade after Brezhnev's death.

Hardin's herdsmen, who ruined their commons, were motivated by the mere logic of private enterprise, which compelled them as rational *homo economicus* to try to maximise their profits. But the Soviet citizens – not only the economic actors – did more than try to maximise bonus. They created a huge illegal counter-economy and indulged in widespread corruption. In both capitalism and 'socialism', plundering the commons is/was the usual practice, but in the USSR it was at a much higher rate. Workers and peasants, who, according to the communists, were the exploited in capitalism but the owners of everything in 'socialism', were also participants in this plunder.

Despite these parallels, there was also a difference, which must be stressed. Hardin's herdsmen were private entrepreneurs; the economic actors in the USSR were not. The motive of profit maximisation is a logically necessary and hence compelling part of the capitalist system. But the motive of bonus maximisation was not a logically necessary part of the 'socialist' system. The working collectives in the latter could have worked sincerely without the bonus motive. The Communist Saturdays were a case in point. In fact, the motive of bonus maximisation was the cause of many evils in the economy of 'socialism'. In Hardin's essay, private entrepreneurs were using a commons. In the USSR, 'socialist' workers and managers working in socially owned enterprises were using the commons in a planned manner. Why didn't it make a difference in the attitude towards the commons?

But let me make another point before discussing the answer. Hardin's pasture and herdsmen are abstract figures, they do not describe an actually existing commons. An actual commons is not a free-for-all thing, it has unwritten rules and understandings, is characterised by a balance of power, consideration and love. In short, it has a moral and environmental order, a commons regime. In contrast, Hardin's commons is rather an open access regime, a free-for-all system without any authority anywhere (cf. *The Ecologist* 1992: 123–30). But this point does not affect our question. The USSR was a sort of huge commons with a sort of commons regime. It had written laws and a functioning centre of authority. Why didn't that make a difference?

The answer that occurs to one immediately is: because the Soviet communists failed to create a new moral order and failed to create and to become, on a mass scale, new men and women, which, they knew, was an indispensable condition for building a 'socialist' society. The egoism of the private entrepreneur and the average citizen of capitalist societies was not overcome in the psyche of the average citizen of the 'socialist' societies. Towards the end, the latter's consumption desires and greed were as high as those of the former, although, for objective reasons, the possibility of fulfilling these desires was much more remote in the 'socialist' countries than in the Western capitalist countries.

According to some observers, it was an impossible project. Harry Nick, a leading economic policy maker of the former GDR, spoke retrospectively of the human factor, of the innate nature of humans, of 'overstraining the moral–ethical motivation'. 'The history of actually existing socialism',[4] he wrote, 'is essentially also a history of repeated disappointments over the non-appearance of the "new man" ...' (Nick 1994: 41). He considered it to have been an impossible project even in respect of leading communists.

> The false understanding of the connection between social ownership [of the means of production] and behaviour of human beings, both the leadership and the people, did not allow any critical or self-critical reflection on the self-interest of the leading stratum. For, no objective basis for power interests could be found where it was searched for, namely in the socio-economic relations where objective causes for human behaviour were supposed to lie. The innate nature of human beings – namely to think mainly of him(her)self and of the immediate purpose – was no subject matter for 'Marxism–Leninism'. (ibid)

What then is the solution? How can the commons be protected? So far as the commons as food basket is concerned, Hardin thought the solution was private property. Goldman also opined that the USSR lacked the first line of defence, because all resources were public property (see chapter 2). This argument for private property is very strong. But no socialist or 'socialist' society could convert its natural resources into private property and still remain socialist in any sense. This is an argument for capitalism, not a means for saving 'socialism' or socialism. The USSR had problems even with the limited legal private enterprise of collectivised peasants.

Moreover, private property is no solution as far as the problem of the commons as cesspool is concerned. Hardin himself wrote:

> The tragedy of the commons as a food basket is averted by private property,

> or something formally like it. But the air and waters surrounding us cannot readily be fenced, and so the tragedy of the commons as a cesspool must be prevented by different means, by coercive laws. … We have not progressed as far with the solution of this problem as we have with the first. Indeed, our particular concept of private property, which deters us from exhausting the positive resources of the earth, favours pollution. The owner of a factory on the bank of a stream – whose property extends to the middle of the stream – often has difficulty seeing why it is not his natural right to muddy the waters flowing past his door. (Hardin 1973:139)

Hardin also admitted that his solution was not just.

> With real estate and other material goods, the alternative we have chosen is the institution of private property coupled with legal inheritance. … We must admit that our legal system of private property plus inheritance is unjust – but we put up with it because we are not convinced, at the moment, that anyone has invented a better system. The alternative of the commons is too horrifying to contemplate. Injustice is preferable to total ruin. (ibid: 145)

As far as pollution is concerned, Hardin knew no other solution than to 'legislate temperance' by means of administrative law. But he raised the question: 'Who shall watch the watchers themselves?' He knew that government inspectors were 'singularly liable to corruption'. But since there was no other solution, he thought: 'The great challenge facing us now is to invent the corrective feedbacks that are needed to keep custodians honest' (ibid: 141). So, even for a strong protagonist of private property, the problem is not solved. The twin practical problems of pollution and corruption and the basic moral problem of justice and injustice remain. Another problem that remains is that of non-renewable resources available to future generations. Hardin has not addressed it at all. Our concept of private property does not deter us from ignoring the claims of future generations on the resources of the Earth, including renewable resources (a pasture is actually a renewable resource). Even a private person can ignore the interests of his/her own grandchildren.

Of course, at the moment, it may appear to some that there is no other solution. But even if we accept Hardin's view that up to now nobody has invented a better system, it does not necessarily mean that a better system cannot be invented in the future. Hardin himself spoke of 'the great challenge' – 'to invent the corrective feedbacks'. But the great challenge is rather to invent and create a *system* in which the commons will not be ruined and justice and honesty will prevail in human relations. I shall discuss this matter in the following chapters.

NOTES

1. According to present-day usage, one should use the word 'human' or 'human being' instead of 'man'. But the 'new man' is an historical designation of an ideal. We cannot change it.

2. Decile ratio means here the ratio of the average pay of the lowest ten per cent of the group (in the pay scale) to that of the highest ten per cent.

3. The Communist Party's youth organisation.

4. That was the term the East European communists used in the 1970s and later to denote their social formation.

CHAPTER 4

The Natural Resource Base of an Economy – Illusions and Realities

Freedom is the recognition of necessity (Hegel)

Any conception of a tolerably good society should be based on sufficient knowledge of the basic material conditions of our existence – those which are given to us as well as those that have been created by us. Among those that are given, some are useful to us, and we may be able to improve their usefulness. Some are constraints, which we may be able to mitigate to some extent, though we cannot wish them away. A conception of a good society should not be merely wishful thinking. The given material conditions, which include our resource base, are of three kinds: the geographic conditions, such as soil fertility, availability of water, temperature, sunshine, and so on; availability of raw materials, for example iron ore, coal, petroleum, wood; and the capacity of nature to absorb the negative impacts of our economic activities, also called its sink function. Many have become aware that we have a problem with our resource base. But there are differences in the conclusions drawn. There are also some who do not see any problem.

In this chapter, I shall discuss some confusions, which need to be removed, and some problematic and controversial points, which need and demand clarity. I shall take it for granted that readers have some general knowledge of the underlying facts.

NON-RENEWABLE RESOURCES

Raw materials

In conventional economic theory, scarcity has always been understood as relative scarcity. With the exception of some free goods (like air), all goods, including raw materials, are considered scarce relative to people's desire to have them. That is why they have a price. But it has always been considered that this scarcity would continuously diminish with capital accumulation and the development of technology and substitutes. So Marx could envisage a society in which everybody would get goods and services according to need. And Keynes could express in 1931,

> the profound conviction that the Economic Problem, ... the problem of want and poverty and the economic struggle between classes and nations, is nothing but a frightful muddle, a transitory and an *unnecessary* one. For the Western World already has the resources and the technique, if we could create the organisation to use them, capable of reducing the Economic Problem, which now absorbs our moral and material energies, to a position of secondary importance. ... the day is not far off when the Economic Problem will take the back seat where it belongs. (Keynes 1931: vii)

But since the early 1970s, we have been hearing of absolute scarcity of resources. One would suppose that at least as far as minerals – which are non-renewable – are concerned, there could be no doubt that they would be exhausted sooner or later. But inveterate optimists denied the existence of any problem. Soon after the publication of *Limits to Growth* in 1972, Professor Wilfred Beckerman of Oxford University said: '... although life on this Earth is very far from perfect there is no reason to think that continued economic growth will make it any worse' (quoted in Georgescu-Roegen 1976: 3), and he was of the opinion that the minerals contained in the top one mile of the Earth's crust would suffice for continuous economic growth for the next 100 million years (Beckerman 1972: 338).

Some historical research in raw materials prices seemed to justify Beckerman's optimism. Barnett and Morse found in 1963 that in the USA, between 1870 and 1957, there had been a falling trend in unit costs in the extractive sector in general and in the mineral sector in particular. This was interpreted as indicating diminishing rather than growing scarcity (cited in Schneiders 1984: 81–2).

Market prices for Raw Materials

In the last twenty years, except in the years immediately after the oil price shocks, world market prices of raw materials, including minerals, have caused no headaches for industrialists and protagonists of continuous economic growth. So Beckerman could reiterate in 1995 his views of 1972 (Beckerman 1995: 56). He showed that since 1972 the 'known reserves' of oil, gas, and several metals have in fact increased, despite the huge consumption that had taken place in the intervening years (ibid: 53).

This apparent contradiction between the logically obvious – that non-renewable resources gradually get depleted – and the facts presented by Barnett & Morse and Beckerman has to be explained. Beckerman has made two interesting points: firstly, he writes:

> There is little point in saying that the increasing use of resources must, *some*

> *day*, come up against finite supplies unless we know roughly when this will happen. How useful is a forecast that it is going to rain *some day*? ... if the limit is to be reached in 100 million years, we might not feel personally involved in the problems of the people alive at that time. ... Tautologies about finite resources, therefore, ... are really not much help in the decision-making process (ibid: 50; emphasis in original).

Secondly, Beckerman points out,

> the usual estimates of known reserves of raw materials are conservative contingency forecasts by the exploration companies and they are related to a certain price: if the price is higher, more resources can be exploited commercially. In other words, the known reserves represent the reserves that have been worth finding, given the price and the prospects of demand and the costs of exploration. ... New reserves are found, on the whole, as they are needed. (ibid: 55–6)

Is there anything wrong in Beckerman's argument? In chapter 2 (p. 28), I quoted a key passage from Abel Aganbegyan's 1988 book. In it he is not saying that the resources of the USSR were exhausted, but differentiating between the more accessible and the less accessible resources, and pointing out the higher costs of extracting the latter. According to the law of conservation of matter, nothing can disappear. When we use the term 'exhausted', we mean only that a resource has become inaccessible or prohibitively costly.

Beckerman's assertion regarding the occurrence of metals in the Earth's crust is both true and useless. The metals are, of course, there – even in ordinary rock. But in what concentration? And how accessible are they? In their book *Beyond the Limits*, Meadows *et al.* (1992) quote the geologist Earl Cook to show the difference between the 'cutoff grade' of metal ores – the lowest concentration of ore that is economically usable – and their average crustal abundance. Cook's table shows that if normal rock is to become a source of metals, the cutoff grade must go down, for most metals, several hundred- to several thousand-fold. Examples are shown in Table 4.1.

'Cutoff grade' is not a fixed figure. It depends on the market price of the mineral and the costs of its extraction; the two are generally related. If the market price rises, sources of the mineral that have earlier been found not economically exploitable, because of too low concentration, can be exploited. But rising prices generally mean that less and less of the resource in question can be used in the production process, and that must have a negative effect on economic growth unless a cheaper substitute is found.

It is logically evident that in the course of time less and less accessible

Table 4.1 Ratios of average abundance and economic viability of some metals

	Average crustal abundance (%)	Mineable cutoff grade (%)	Ratio (:1)
mercury	0.0000089	0.1	11200
lead	0.0012	4.0	3300
zinc	0.0094	3.5	370
copper	0.0063	0.35	56
iron	5.820	20	3.4
aluminium	8.3	18.5	2.2

Source: Meadows *et al.* 1992: 84–5.

resources and minerals with fewer and fewer degrees of concentration have to be used and that this, normally, should lead to a rise in the costs of extraction and hence in prices. Aganbegyan's explanation of the former USSR's problems with resources is therefore convincing. But this logic applies everywhere. Why, then, has there been no problem of resources in the West?

One set of reasons has already been stated, namely the West's control over and easier access to the resources of the whole third world (an advantage that the USSR did not enjoy), the third world's ever worsening terms of trade *vis-à-vis* the first world, its ever worsening external debt problem, and finally its rock-bottom wages.

As for resources extracted in the Western industrialised countries themselves, since 1957 – the cutoff date of Barnett and Morse's study – unit costs have been rising. For example, in West Germany the cost of extracting bituminous coal has long been rising because one has to dig deeper and deeper for it. In 1997, it cost DM280 per ton, compared with DM80 per ton of imported coal (*Frankfurter Rundschau*, 12 March 1997). F.E. Trainer collected enough evidence from the 1960s to the end of the 1970s to be able to write in 1985 that in the USA 'evidence indicates that mining investment is yielding sharply diminishing returns' (Trainer 1985: 51).

However, in spite of generally diminishing returns, and in spite of generally low costs of extraction in the third world, the extractive industries have not come to a standstill in the industrialised countries of the West. For instance, in the USA, before 1910, only copper ore containing over 2 per cent of the metal was extracted. Since then, copper ore containing a smaller and smaller percentage of the metal is being used for production. Copper ore extraction in the USA has not stopped (Meadows *et al.* 1992: 86). This has to be explained.

Firstly, there is a difference between Aganbegyan's report and that of Barnett and Morse. Whereas Aganbegyan speaks not only of costs, but also of actual deposits, Barnett and Morse speak only of unit costs. Volker Schneiders remarks: 'Theoretically, unit costs today can be lower than they were yesterday, while tomorrow a mine can be exhausted' (Schneiders 1984: 83). Secondly, unit costs depend on many factors: accessibility, labour intensity, wages, technology, price of energy and of machines, and so on. Cheap energy has long been playing an increasing role in all industries and rapidly replacing costly labour.

So now we come to the other reason why the West did not have a resource problem. Generally, in the extractive industries, increasingly difficult geological and other natural conditions hamper extraction. But in the West's sphere of influence, in the case of energy, especially oil – a central and universal means of production in the extractive industries – an exactly opposite development took place in recent decades (ibid: 84–5). Easily extractable, cheap and profusely flowing oil around the Persian Gulf and the Caribbean Sea permitted unit costs of other resources in the West not to rise (too much) or even to fall. Obviously, in the USSR the average natural productivity in the energy sector could not compensate for the growing difficulties in the rest of its extractive sector – lack of cheap labour, lack of neo-colonies, and mismanagement.

The conclusion that emerges from the above is as follows: if energy is cheap and profuse, it might be economical to reach a depth of one mile into the Earth's crust and extract raw materials from even very low-grade ores. Then there would be no scarcity. This has been formulated as a general belief, which Georgescu-Roegen called 'the energetic dogma': 'All we need to do is to add sufficient energy to the system and we can obtain whatever material we desire' (Brown *et al.* quoted in Georgescu-Roegen 1978: 17).

Whether cheap and profusely flowing energy will be available in future is a question I shall deal with soon. Let us, for the sake of argument, suppose that energy will continue to be available at the same price and in the same quantity as today. Even then, we know that progressively lower grades of ores are being processed to obtain metals. The consequences can be estimated from the exponentially growing amount of rock that has to be crushed, transported, processed, smelted, and removed to waste dumps. F. Bender calculated: 'for obtaining one ton of copper (0.6% copper in ore) today 200 tons of ore have to be mined. … If we want to use rock with the average composition of the Earth's crust as copper ore, then we need 25,000 tons of rock for one ton of copper …' (quoted in Schneiders 1984: 86). And

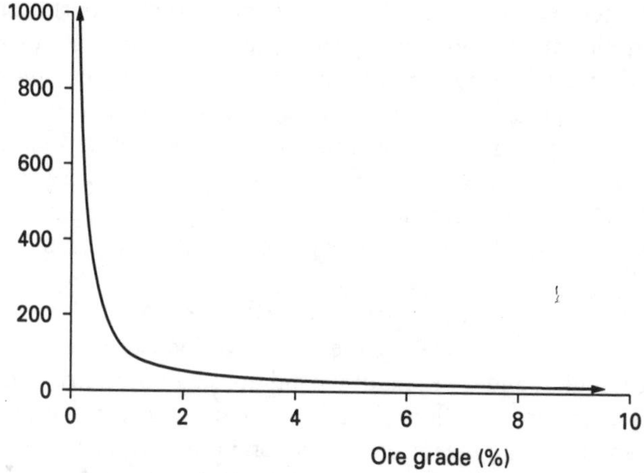

Figure 4.1 Tons of waste per ton of metal

> for instance, in order to obtain one ton of copper from ore with copper content of 0.5%, the amount of energy necessary ... is 0.72×10^{11} Joule. This energy requirement grows hyperbolically with the arithmetical diminution of metal content in ore. Thus, the theoretically required amount of energy for obtaining copper from normal rock from the Earth's crust with an average copper content of 50 ppm, is 2.16×10^{12} J/t (Joule per ton). (ibid)

The above estimates can be illustrated with the help of the graphs given in Figures 4.1 and 4.2.

As the average grade of ore declines through depletion from 8% or more to 3%, there is a barely perceptible increase in the amount of mining waste generated per ton of final metal. Below 3%, wastes per ton increase dramatically. Eventually the cost of dealing with the wastes will exceed the value of the metal produced (Meadows *et al.* 1992: 87).

As their metal content declines, ores require increasingly large amounts of energy for their purification. (Source: N.J. Page and S.C. Creasey, quoted in Meadows *et al.* 1992: 116).

I think these facts and estimates permit us to say that Beckerman's optimism is only indulging in an illusion.

Energy

The 'energetic dogma' is theoretically quite plausible. With abundant cheap energy we may even be able to change geographic conditions, for example, convert sea water into fresh water and use it to make

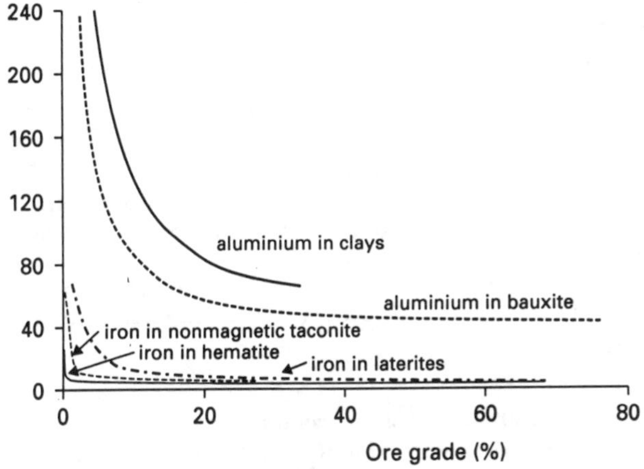

Figure 4.2 Thousand kilowatt-hours per ton of metal

deserts bloom. We may then be able to produce rain artificially. The question is, however: how good are the prospects of getting abundant cheap energy in future?

Fossil fuels

For a theoretical book like this, the exact number of years that known and presumed reserves of the fossil fuels would last is unimportant. Nevertheless, the estimated figures are given in Table 4.2.

> Do these rising reserve/production ratios mean that there were more fossil fuels to power the human economy in 1990 than there were in 1970?
>
> No, of course not. There were 450 billion *fewer* barrels of oil, 90 billion *fewer* tons of coal, and 1100 trillion *fewer* cubic meters of natural gas. Fossil fuels are non-renewable resources. When they are burned they … do not, on any time scale of interest to humanity, come back together to form fossil fuels again. …
>
> Those who see the discoveries of the past twenty years as proof that there are no limits to fossil fuels are looking at only part of the energy system. (Meadows *et al.* 1992: 68)

And these resources are becoming less and less accessible; hence – and that is the main point – they will become ever dearer.

As far as global prospects for oil are concerned, the summary of the current prognoses of various experts and institutes are as follows:

Table 4.2 Annual production and reserve/production ratios for oil, coal, and gas, 1970 and 1989

Fuel	1970 production (per year)	1970 R/P (years)	1989 production (per year)	1989 R/P (years)
Oil	16.7 billion barrels	31	21.4 billion barrels	41
Coal	2.2 billion tons	2300	5.2 billion tons	326 (hard coal) 434 (soft coal)
Gas	30 trillion cu ft	38	68 trillion cu ft	60

(Meadows *et al.* 1992: 68)

> Today, in about 90 per cent of the world's deposits, oil can still be extracted at a cost below US$ 8 per barrel. But the USA's reserves will last for only another eight years, those in the North Sea for another 17 years. After that, new exploration will be costly. Extraction costs in the USA will be US$15–20 per barrel, in Europe US$12–15. By the year 2010, according to the Prognos Institute, extraction costs in non-OPEC states will rise to US$20–30 per barrel. In the long run, only OPEC can increase its supply without any problem and at favourable costs. ... At the latest after the year 2000, rising oil prices will again dominate the scene. ... US$26 appears to be quite realistic for the year 2010. (Müller 1993: 67–8)

Trainer refers to Petroconsultants' 1995 report, *World Oil Supply 1930–2050*, which concludes that world production and supply will probably peak as early as the year 2000 and will decline to only half the peak level by 2025. The report predicts large and permanent increases in oil prices soon after peak production (Trainer 1996: 8).

As regards oil extraction in the USA, after 1940 prospecting led to fewer and fewer finds, which indicated that reserves were coming to an end. After the maximum, achieved around 1970, oil extraction in the USA (excluding Alaska) fell by 40 per cent. The find in Alaska could raise the total figure again, but not back to the 1970 level (Meadows *et al.* 1992: 72). Moreover, the cost of extracting oil in Alaska is higher than elsewhere in the USA.

Globally, we consume 23 billion barrels of oil a year, but we are discovering only 7 billion barrels a year (Trainer 1996: 8). The upward revisions in 'known reserves' in the recent past were not all due to new discoveries. Some existing fields were declared to be bigger than originally stated. But, writes Trainer,

> As an oil field ages and becomes more depleted, increasing amounts of energy have to be used to pump the oil out. ... pressurised water, etc., have

> to be pumped down. ... These processes take a lot of energy and this has to be subtracted from the amount of energy in the oil retrieved. ... evidence indicates that for the USA the two curves will actually meet around 2005. ... That is, it will then not pay to produce oil, because to find and pump a barrel will take more energy than there will be in the barrel. ... at some point in time we will have to stop producing even though a lot of energy is still left in the deposit. (ibid)

The oil industry seems to be well aware of the situation. No new super-tankers have been built for 20 years.

The number of years a non-renewable resource would last is usually calculated by dividing the known reserve by *current* production or demand. But today, the average individual in the greater part of the world population is not consuming as much of these resources as an average North American or Central European. It is not as if the average third world citizen does not need as much oil as the average German. What prevents him or her from demanding more is that these resources are already too costly. Beckerman's optimism – 'new reserves are found, on the whole, as they are needed' – is therefore nonsensical. It does not help the poor if Saudi Arabia has a lot of oil and Siberia a lot of gas, if they are too costly for the poor to buy. India direly needs oil reserves of her own, but despite intensive prospecting, little has been found in her large territory.

Nuclear Energy

Nuclear (fission) energy is also not inexhaustible, since the fuels that nuclear reactors use are also non-renewable. But that is not my point here. Nor is it my point that nuclear energy, especially the fast breeder reactors and plutonium, are very dangerous. My point is that nuclear energy will never become cheap. It is well known that many US electricity companies have in recent years cancelled plans to build nuclear power plants because generating nuclear electricity has become costlier than generating electricity from coal. Moreover, it has been estimated by the Prognos Institute that if nuclear power plants have to be fully insured against all possible damage, then nuclear (fission) energy would cost about DM6 (c. US$3.43) per kWh (quoted in Alt 1993: 7).

Thermonuclear energy from fusion reactors has been considered by many for the last 30 years or more to be the final answer to humanity's energy problems. In theory it is convincing. The raw material, deuterium, is available in almost inexhaustible quantities in the oceans. The fusion reactors are said to be much safer than the fission reactors. And emission of radioactive substances from them is expected to be much less. In short, if it becomes reality, enormous amounts of energy could be available for all times.

But at what price? There are many contradictory reports on this question. From the latest reports it appears that another 50 years of research and experimentation will be necessary before thermonuclear energy can be used for economic purposes. And many more billions of dollars would have to be invested in research. The costs thereof have already become so heavy that even such rich countries as Germany and France have declared their lack of interest in having the planned joint-venture International Thermonuclear Experimental Reactor (ITER) located in their territory. For, according to the agreement, the host country has to bear 70 per cent of the costs. In the meantime, the USA has cut its fusion research programme by one-third (*Frankfurter Rundschau*, 18 December 1993; 23 September 1996).

There is another problem. A thermonuclear reactor using only deuterium would be much less effective than one using a mixture in equal parts of deuterium and tritium. But tritium is not available from natural sources. It is produced through nuclear reaction between lithium-6 and neutrons. Lithium-6 is a scarce metal – exhaustible.

All this information indicates that thermonuclear energy, if and when it became available, would not be cheap. Professor Heindler of the University of Graz warns against too great hopes. According to him, this kind of energy would be only 'a small extra' added to the energy mix of the future (*Frankfurter Rundschau*, 23 September 1996).

To sum up, the sources of energy discussed so far would be neither cheap nor abundant. On the contrary, they would become ever scarcer and ever more costly. They would, therefore, not be available to pierce the Earth's crust to a depth of one mile to obtain resources from low-grade deposits and then make them generally available.

RENEWABLE RESOURCES

Among activists of the ecology movement one often hears euphoric statements on renewable resources. Franz Alt, an environmentalist and popular German TV journalist, writes:

> Just solar radiation contains about 10,000 times as much energy as present-day world energy consumption; the winds 35 times as much; the growth of biomass 10 times as much; and hydro-power offers half as much. …
>
> Reed grass that can be grown on the fields that today lie fallow in Germany can produce as much energy as all the 21 nuclear power plants of the country. …
>
> The working group 'Solar Energy for Environment and Development'

> of the United Nations ascertained at its conference held in March 1991 in Rome: 'It is certain that the total potential of renewable energies is about 10,000 times the current total energy consumption of humanity. (Alt 1993: 6–8)

Alt also asserts that from biomass we could get raw materials for almost everything: houses, cars, every kind of chemical, and so on. And all such materials could finally be composted (ibid.).

Let us now examine these and other similar assertions.

Solar energy – the question of energy balance

The great advantages of solar energy are obvious and well known. So why hasn't it yet replaced all the forms of energy that are problematic? The main problem is that modern humans want to have energy mainly in the forms of electricity and liquid fuel (and to some extent gas), for these are the most useful. The warmth we get from sunshine without any additional effort is also energy. It is very useful for agriculture, drying wet linen, and making us feel comfortable in winter. In industrial processes, however, high-intensity heat is needed. This cannot be had from the sun without some additional effort – not even warm water in the bathroom. The additional effort necessary to get solar energy in the desired form makes it very costly.

The most desired form is solar electricity. Today, in Germany, the production cost of 1 kWh of solar electricity is DM2, roughly twenty times that of conventional electricity. In the tropical deserts, where the sun shines every day and more intensely, or in sunny southern California, it is much less. Christopher Flavin of the Worldwatch Institute claims that in 1993, production of 1 kWh of solar electricity cost 25–40 cents (in January 1997, US$1 = DM1.60), but he does not reveal where and how this value was arrived at. Nevertheless, he says that solar electricity is 'still too expensive to compete head-to-head with conventional generating technologies', and 'prices will need to be cut by a factor of three to five before large-scale grid-connected applications become economical' (Flavin 1995: 66–9). Solar electricity enthusiasts say it could soon become as cheap as conventional electricity; it only needs to be used on a large scale, so that producers of collectors[1] can enjoy economies of scale.

Meadows *et al.* envisage solar energy one day replacing fossil fuels. Paraphrasing a 'simple rule' suggested by Herman Daly, they write:

> For a non-renewable resource – fossil fuel, high-grade mineral ore, fossil groundwater – the sustainable rate of use can be no greater than the rate at

> which a renewable resource, used sustainably, can be substituted for it. (For example, an oil deposit would be used sustainably if part of the profits from it were systematically invested in solar collectors or in tree planting, so that when the oil is gone, an equivalent stream of renewable energy is still available.) (Meadows *et al.* 1992: 46)

It would be wonderful if this vision could become reality. Then we would have solved all our energy problems and much of our pollution and global warming problems. But in the case of no other resource is the discrepancy between hope and reality as great as in that of solar energy. In 1976 Barry Commoner wrote: '... solar energy can not only replace a good deal, and eventually all, of the present consumption of conventional fuels – and eliminate that much environmental pollution – but can also reverse the trend toward escalating energy costs that is so seriously affecting the economic system' (Commoner 1976: 122). He visualised liquid hydrogen produced from water by using solar electricity as the storage medium of solar energy.

As stated above, the 1996 production cost of 1 kWh of solar electricity in Germany was DM2. According to the Information Centre of the Electricity Industry (of Germany), if this electricity is converted into liquid hydrogen, the fuel would be 50 times as costly as conventional sources of energy (*Frankfurter Rundschau*, 23 March 1990). This may be an exaggeration by an interest group opposed to the new fuel. But according to Trainer, who is not at all opposed to alternative sources of energy, 50 per cent of the energy would be lost in the process of conversion. And if solar electricity were produced in the Sahara desert, where the sun shines every day, converted into liquid hydrogen, transported to Europe, and then converted back into electricity, 95 per cent of the energy originally collected would be lost (Trainer 1996: 9).

The problem is ultimately not that of costs and prices. If it were, the state, by imposing some tax, could make energy from fossil fuels and uranium as costly as solar energy, and so make the latter competitive. The problem is more serious. The life-span of a solar-power plant is at the most 20 years. What would happen thereafter, when the solar collectors and other equipment would have to be replaced? The sun, of course, will still be shining. But the materials required to build and operate solar-power plants are mostly exhaustible. And, if the idea of Meadows *et al.* were implemented, the oil fields that have financed the first generation of solar-power plants would also be empty by then.

Can the second generation of solar-power plants be built using the solar energy produced by the first generation? Georgescu-Roegen's answer is: no, at least not yet. He differentiates between 'feasible' and 'viable' technologies. A viable technology is one that is capable of *reproducing itself* after it has been brought into existence by means of an earlier technology. Illustrating the point, he writes: 'The first bronze hammer ... was produced by some stone hammers. However, from that moment on, all bronze hammers were hammered only by bronze hammers' (Georgescu-Roegen 1978: 18). To take another illustration, the first coal was extracted by using human and animal muscle power. But soon, machines driven by coal energy were producing the capital equipment necessary to extract coal, and such equipment was itself to be driven by coal energy. This is not the case with solar energy. All the necessary equipment, including solar collectors, are produced through processes based on sources of energy other than the sun (coal, oil, uranium, etc.). Solar energy is, therefore, feasible only so long as other sources of energy are available. That means it is not viable.

The reason for this state of affairs is, according to Georgescu-Roegen, that all known recipes for the direct use of solar energy produce a deficit in the general energy balance, that is, they indirectly consume more of some other form of energy than they produce directly. It may be impossible to solve the problem, for the intensity of solar radiation reaching ground level is extremely low. Moreover, the sun does not shine at night. These two facts are cosmological constants beyond our control (ibid: *passim*). According to Georgescu-Roegen, 'the fact that, in spite of the present din about solar energy, not even experimental pilot plants' that produce collectors and other necessary capital equipment with the aid of some solar energy 'have been set up by one of the various R & D agencies working in this direction, is sufficient proof that a technology [for producing energy] based on collectors is not viable' (ibid: 19).

Georgescu-Roegen anticipated the argument that it is only because of prevailing market prices that collectors and other connected capital equipment are not produced with the aid of solar energy, but with the aid of fossil and nuclear energy (ibid). This is convincing. It would be nonsensical to produce anything with the aid of solar electricity as long as its price is higher than that of conventional electricity. But that also means that if the second generation of solar power plants were built with the aid of solar electricity costing DM2 per kWh (rather than conventional electricity costing about DM0.10 per kWh), the electricity produced by them would cost more than DM2 per kWh, and the third generation still more, and so on.

It is easy to understand that no business can function if the output is less than the input. If we produce with the aid of electricity a different product, say mobility or light, then a person who desires mobility or light may be willing to pay any price for electricity. But if we produce merely electricity with the aid of electricity, it makes sense only if output exceeds input. But the whole thing is difficult to understand because solar energy enthusiasts have confused the general public with their unsound assertions in regard to the energy balance of solar electricity. Thus in 1982 one could read in a book written by a group of alternative energy experts from Tübingen:

> Of course, ... until recently, monocrystalline solar cells supplied only in 13 years as much energy as required for their production. But already today, polycrystalline solar cells have an energetic amortisation time of less than five years. But since they are today still too costly, we have not included solar cells at all in our scenario. ...' (Arbeitskreis Alternativenergie Tübingen 1982: 178)

In a much-quoted paper published in 1991, Palz and Zibetta asserted that in European climates, the average 'energy pay-back time' (i.e., energetic amortisation time) for photovoltaic (PV) amorphous silicon modules and crystalline silicon modules (they examined polycrystalline modules) were 1.2 years and 2.1 years respectively (Palz & Zibetta 1991: 211–16). They concluded: 'The energy pay-back times of current photovoltaic technology are by and large comparable to that of large-scale electricity production in fossil and nuclear power plants' (ibid: 215). But in 1995, in a well-researched paper, Christian Friedl informed us that the energy pay-back time for a PV module used in Düsseldorf was 9 years, and he made the additional comment that 'polycrystalline solar modules have a better ecological balance because the energy-intensive pulling process of the monocrystal is dispensed with' (Friedl 1995: 34). Flavin, in his otherwise thoroughly researched paper, does not go into the question at all.

Several questions arise immediately after reading these assertions – questions to which no answers have yet been given by solar energy enthusiasts. (I have the feeling that they avoid such questions.) Firstly, the big difference in reported energy pay-back times – 1.2 years and 9 years – must be explained. Secondly, if polycrystalline PV cells are so superior to monocrystalline ones, why are the latter being produced at all? Thirdly, if polycrystalline cells need much less energy for their production, why are they too costly, as the Tübingen group writes? Fourthly, if the energy pay-back times of current PV technology are as low as those of fossil-fuel and nuclear power plants, why is the production cost of solar electricity in Germany roughly 20 times that of conventional electricity? Why is it still not competitive?

The argument of solar electricity enthusiasts, that PV cells need only to be produced on a large enough scale in order to become competitive, is not understandable. After all, PV cells are produced in factories with modern machines, whereas thermal and nuclear power plants and dams for hydroelectricity are not produced on a large scale in factories. Construction of these is surely a much more labour-intensive process than the factory production of PV cells and other collectors.

It seems to me that the answers to these questions, and the explanation of the contradictions and anomalies, lie mainly in the obviously incorrect way that energy pay-back times have been calculated. Failure to add up *all* the expenditures of energy involved perhaps explains the big differences between the results.

A careful reading of Friedl's paper and that of Palz and Zibetta shows that they have begun the calculation only at the stage where silicon dioxide is smelted to produce silicon. Palz and Zibetta have, moreover, assumed that the energy content of the silicon used is 20 kWh/kg, which is the energy content of ordinary metallurgical grade silicon, whereas the silicon used in the PV industry comes actually from residues of the electronic industry, where the energy content of silicon is 200 kWh/kg. They argue that 'the energy content of a rejected material ... is ... debatable'. But they themselves write that the PV industry needs this 'rejected' silicon (and not ordinary metallurgical grade silicon) because although 'its purity is much lower than that of electronic grade silicon ... [it] is still high enough for PV technology' (Palz & Zibetta 1991: 212).

But the chief criticism against their method is that one cannot begin the calculation somewhere in the middle. Georgescu-Roegen demands, convincingly, that in the case of, say, aluminium mirror collectors, one 'must include the mining of bauxite, its transportation, as well as the reproduction of the entire plant for extraction and reduction of alumina' (Georgescu-Roegen 1978: 19). In the case of PV cells, one must include the mining of silicon dioxide, its transportation, the reproduction of the plant, and so on. This has not been done by solar energy enthusiasts. If one implements Georgescu-Roegen's demand, one would easily come to the conclusion that the energy balance of solar electricity production is negative: that it is, therefore, feasible but not viable.

A similar confusion apparently prevails in the case of thermonuclear energy. In 1988, it was reported that in the Livermore laboratories (USA), a controlled nuclear fusion produced more energy than that needed for the initial ignition (*Frankfurter Rundschau*, 26 August 1988). But five years later another report stated that the fusion process still needed eight times as much energy as it produced (*Frankfurter Rundschau*, 18 December 1993). In both cases, it appears that the energy required to collect deuterium from (sea) water, to produce tritium from lithium-6, and to reproduce the plant had not been included in the calculation. But such costs exist. If after 50 years thermonuclear energy for economic use becomes feasible, it might well not be viable.

A somewhat similar doubt exists about the energy balance of nuclear fission. Hardin B.C. Tibbs, no opponent of industrial society, writes: 'Even nuclear fission, viewed over the long term, probably suffers from a low or even negative net energy yield when the total "life-cycle" cost of construction, fuel production, decontamination, decommissioning, and waste storage is deducted' (Tibbs 1992: 16).

A theoretical point might be of interest in this matter. It is clear that there is a difference between energy balance and financial balance. While calculating the energy balance of any technology for producing energy, one includes the energy consumed and excludes human labour,[2] which, however, has financial costs. So if a technology could replace energy by human labour, then its energy balance would look better, but its financial balance may, especially in high-wage countries, become worse. As far as solar electricity is concerned, its present high cost of production might be reduced to some extent if its labour-intensity could be reduced through the use of more automatic machines. But that would increase the expenditure of energy and worsen the energy balance. You cannot improve both balances.

Large-scale production might to some extent help reduce both labour and energy cost per unit. But there are limits to economies of scale. There are even diseconomies of scale. Moreover, as shown above and in chapter 2, there are other factors that, in the long run, cause a general and continuous increase in costs. If the price of conventional energy, which is used for manufacturing collectors and other necessary equipment rises, that will also generate an upward pressure on the price of solar energy. Increasing the scale of production cannot alone cancel out the effects of such factors. Cars are being produced in ever increasing numbers, but they are not becoming cheaper.

Very often one hears the argument that large-scale production would make the price of PV cells tumble as it did in the case of computer chips. High price and low price are relative terms. In fact, the price of PV cells *has* tumbled since they were invented in 1954. The reasons why computers have become highly economical while PV modules, generally, have not yet, are as follows: firstly, whereas computers and chips have become smaller and smaller, one still needs to cover 2.5 sq km with PV modules to produce 200 MW of solar electricity; secondly, whereas computers provide a service, namely data processing, for which we earlier paid a very much higher price, PV modules produce only electricity, a banal product, for which we are paying a much lower price than the cost of production of PV electricity. In a somewhat similar way, there is a difference between cars and computers. Whereas a car must be able to move five people and luggage, a computer must handle only intangible things: electrons carrying information expressed in monochrome symbols.

The argument that strong vested interests are holding back PV technology is not convincing. The producers of wet razors could not hold back the electric shaver, and the mighty coal industry of Great Britain could not hold back nuclear power, nor prevent the import of cheaper coal. The same industrialists can, moreover, produce coal-fired or nuclear or solar power. The only consideration for them is profitability. No business can run at a loss. And no state can go on giving subsidies without end to a technology that produces one unit of energy at the cost of, say, two units of energy.

To go a little deeper in theory, producing energy from sunshine has a special disadvantage compared to producing it from fossil fuels. The latter is available to us in low-entropy state, which in fact is the result of the concentration and storage of solar energy by nature over millions of years. But sunshine reaches us on Earth in a high-entropy state. The second law of thermodynamics requires us to use low-entropy energy from outside (coal, oil, etc.) to convert the high-entropy energy of sunshine to low-entropy energy. In other words, we must ourselves collect diffuse sunshine, convert it to electricity or heat and store it through elaborate artificial processes. Of course, coal and oil also have to be collected. But the high concentration of energy in them makes it possible to collect them without requiring energy from outside.

This point is often found difficult to understand. Even Barry Commoner made the mistake of comparing sunshine with rainfall. He wrote:

> Like sunlight, the energy of falling rain is widely diffused across the earth, and its gentle force would seem to hold no promise of delivering power sufficient to run the energy-hungry tasks of modern society. ... What transforms the diffuse 'impractical' energy of the rain into the eminently useful power of the hydroelectric plant is the process of concentration. (Commoner 1976: 132)

Commoner, astonishingly, fails to notice the difference that diffusely falling rainwater collects *itself* behind a dam because of natural topography, whereas diffuse sunshine has to be collected *by us* by covering a large area with very costly solar modules or mirrors.

Summing up, Georgescu-Roegen writes:

> The truth is that any present recipe for the direct use of solar energy is a 'parasite', as it were, of the current technology, based mainly on fossil fuels. ... And it goes without saying that, like all parasites, any solar technology based on the present feasible recipes would subsist only as long as its 'host' survives. (Georgescu-Roegen 1978: 19)

This was written in 1978. Fourteen years later, in spite of all the improvements in PV technology, Georgescu-Roegen saw no occasion to change his opinion (1992: 16–18).

Solar electricity has become a success notably in remote sunny areas unconnected to the national power grid, where the cost of such a connection is about as much as that of a PV module sufficient for the given purpose (Flavin mentions vacation cabins in the USA). But Flavin's hope that 'over time, solar- and wind-derived hydrogen could become the foundation of a new global energy economy' (Flavin 1995: 72) is an illusion. Perhaps some day a more efficient technology will emerge. But any energy technology that wants to succeed the present ones must first pass Georgescu-Roegen's viability test.

Biomass

It is of course nonsensical to say that solar collectors financed through the profits from an oil deposit would, when the oil is gone, supply an equivalent stream of renewable energy. But Meadows *et al.* (1992: 46) mentioned tree planting as an alternative. In fact, Herman Daly's paper, to which they referred, contains no mention of solar collectors. Daly spoke only of 'tree planting for wood alcohol' (Daly 1990: 4). This is more convincing. Wood may become a substitute for fossil fuels. After all, from time immemorial wood has served as our main source of energy, and it is renewable.

Actually, rather than just wood, we should consider biomass in general as the source of renewable energy. In the recent past, many other plants and plant products have been suggested as alternative sources of energy and raw materials. One may ask why it is necessary to consider them if they have already proved their qualities as renewable resources. The necessity arises from the fact that modern industrial economies want energy mainly in two forms: electricity and liquid fuel, and to a lesser extent as combustible gas. It was not so before the industrial revolution.

Renewable energies are, in principle, inexhaustible, but they are not necessarily cheap and abundant. In the past, construction of windmills and watermills was costly. Horses had to be fed with oats, which had to be grown with the expenditure of much human labour. Trees had to be felled to get wood. But trees and other plants are superior to solar collectors. They concentrate and store solar energy over a long period through a natural process. However, if we use them as our main source of energy, we have to accept their naturally slow tempo of growth and devote much land to them.

I have no data on wood alcohol, but enough on several other biological sources of energy and raw materials. Bioethanol can, in principle, be obtained from any plant product containing sugar or starch, and can be used as a complete substitute for or added to petrol. In Europe, producing what is called biodiesel from rape seed is being promoted by the European Union. Both of these fuels can be used for motor vehicles.

But one should not think that all environmental and resource problems of industrial societies could be solved with these renewable resources. As in the case of direct use of solar energy, there is some confusion as regards the energy balance of biofuels. While an EU commissar made the general claim that research had shown that the energy balance of biofuels was positive, both governmental and non-governmental agencies disputed this claim (*Frankfurter Rundschau*, 20 February 1992).

According to the Federal Environment Office of Germany, the energy balance of bioethanol from sugar-beet, wheat, and potato is negative (*Die Tageszeitung*, 1 April 1992). The European Environment Bureau (established by several associations for environmental protection) stated that the energy balance of bioethanol is barely positive if the by-product straw is included in the calculation (*Frankfurter Rundschau*, 20 February 1992). Although the energy balance of rape seed oil is generally considered to be positive, there is doubt about biodiesel, the production of which from rape seed oil needs further energy. The reason for the energy deficit of these biofuels is that chemical fertilisers and pesticides are needed to grow the crops. The energy content of such chemicals constitute, for example, 40–60 per cent of the energy content of rape seed (Schmitz-Schlang 1995: 14).

As far as cost of production per litre is concerned, whereas for petrol it is only DM0.30, for bioethanol it is DM1.20–1.60 (*Die Tageszeitung*, 1 April 1992). In the German Federal Ministry of Agriculture, it was estimated that production of this fuel from sugar-beet, if it were to be promoted, would need an annual subsidy of DM4,519 per hectare, from wheat DM2,406 (Schmitz-Schlang 1995: 23). For biodiesel, which is being promoted, the annual subsidy – in addition to the indirect subsidy given in the form of exemption from the mineral oil tax – is DM1,333 per hectare (ibid.).

In Brazil, the state had to promote the bioethanol industry with tax and finance concessions, and guaranteed low prices (probably through subsidies) (Göricke & Reimann 1982: 330–1). I have no data on the energy balance of bioethanol produced from sugar-cane in Brazil, but Brazilians have certainly not spent much petroleum energy to produce bioethanol. The whole purpose of the programme was to reduce petroleum imports. If they are using labour-intensive methods, then it is quite possible that the energy balance of Brazilian bioethanol is positive. It may be even more positive if one includes the energy obtained from bagasse burning in the calculation.

Any kind of biomass can be burnt to produce heat. However, only dry, woody biomass is suitable for this purpose. Wet biomass is more suitable for biogas production. Straw is a known source of energy, which can conveniently be burnt directly in the farm. But if transported to distant places, it has to be compressed into briquettes or pellets; this is a costly process. A straw briquette is about as costly as a lignite briquette (Michelsen & Öko-Institut 1991: 257). Straw is much more useful in agriculture itself. In future, if organic farming becomes the rule, much of it must be returned to the field in order to maintain the humus content of the soil.

Some kinds of fast-growing grass – especially elephant grass, reed grass, and cambric grass – are also being discussed and experimented with. Their energy balance can be positive if harvested in a dry state. But in temperate zones, where it rains throughout the year, this cannot be guaranteed. These are tropical plants, which do not grow in temperate-zone winters. In Germany, discussion of them has died down.

It appears at present that – apart from burning, and perhaps also gasifying, wood – biogas from all kinds of biological wastes, including human and animal excrement, is the only form of energy from biomass that is without doubt sensible. Biogas has been extensively produced in underdeveloped countries such as India and China, so that its energy balance and cost of production must be assumed to be favourable. Even in Germany, in the 1950s, a number of biogas plants were functioning satisfactorily (IRB 1985: 17). Moreover, the residues make good quality manure. But biogas may not be the right fuel for motor vehicles, as many eco-enthusiasts think. The energy and financial cost of liquifying it is very high. Burning it to produce heat and perhaps also electricity appears to be more viable.

But cultivating grass and other plants on agricultural land for energy production in the form of biogas, or any other form, is problematic. It depends on whether surplus land is available after meeting the food needs of humans and domestic animals. This depends on the size of the population and the kinds of food needs. If extensive organic farming has to be practised in future, then more land will be needed for food production than today. Friends of the Earth Netherlands have excluded such things from their scenario for the year 2010, *Sustainable Netherlands*, because 'it would take up far too much arable land' (Brakel & Zagema 1994: 25). Moreover, we must respect the land needs of the other species of the planet.

Biomass can also be, indeed it already is, a source of raw materials for many purposes. Apart from the well-known raw materials of organic origin, the chemical industry is carrying out research to replace petroleum as the raw materials base for products such as plastics and colours by biomass. Starch, oils, fats, ethanol, glucose, phenol, glycerine, etc., have been mentioned as possible products. But the production costs are still too high. Perhaps they will be marketable when petroleum becomes too costly.

Environmental problems

Many eco-enthusiasts imagine energy and raw-materials production from biomass to be free from environmental problems, on the grounds that what nature builds up, it also degrades naturally. But the matter is not as simple as that.

One thing is certain: CO_2 concentration in the atmosphere cannot increase through the use of biomass itself. Any form of biomass generates through burning only as much CO_2 as it has neutralised in the process of growing. In other respects, energy and raw materials production from biomass is far from being environmentally benign. Since, for crops from which biodiesel and bioethanol are produced, only intensive methods of cultivation are being practised or considered, large amounts of chemical fertilisers and pesticides need to be used. And that has well-known negative effects: nitrates and pesticides in ground water, and emission of the trace gas nitrous oxide (N_2O), which contributes to the greenhouse effect 300 times more than CO_2. Intensive agriculture also contributes to soil erosion and the diminution of biodiversity (Schmitz-Schlang 1995: 26; *Die Tageszeitung*, 1 April 1992).

There are also problems with the waste water of such production. In the case of bioethanol production from sugar-cane, large quantities of the effluent vinasse (waste water with residues) flow into the rivers, killing fish. In Brazil, in the early 1980s, several rivers and sea coasts were rendered biologically dead (Göricke & Reimann 1982: 334). In the case of fast-growing grasses and trees, the environmental impact is negligible, because they are never (recommended to be) grown with chemical fertilisers or pesticides. That is also why one may safely assert that their energy balance can be quite positive. (Eucalyptus is problematic, but for different reasons.)

Hydroelectricity

Hydroelectricity is cheap, but not just any site on a river is suitable for its generation. Experts think that in Germany the scope for large and medium-sized hydro-power plants is exhausted, although only about 5 per cent of the country's electricity comes from this source. If the potential for small hydro-power plants were fully realised, this figure could be doubled (Michelsen & Öko-Institut 1991: 256).

The most serious problem with hydroelectricity is the silting up of the reservoirs. In temperate zones, the weathering of rocks, the first step in the creation of silt, is slow. But in the tropics, the process is much faster owing to high temperatures and heavy rains. Moreover, chemical weathering of rocks is 20–50 times faster in the tropics than in temperate zones (Pearce 1992: 228).

The Anchicaya reservoir in the Colombian Amazon silted up within 12 years of coming into operation. The reservoir of the Tarbela Dam on the Indus in Pakistan is filling up with silt at a rate approaching 2 per cent per annum. The reservoir of the Nizamsagar Dam in central India lost two-thirds of its capacity to silt within 30 years. Dams like Tarbela are now estimated to have a lifetime of 40 years or less. Dozens of major dams are likely to become useless by early in the 21st century (ibid: 228–9). Summing up, Pearce writes: 'Many of the largest hydroelectric dams in the world today, occupying the best sites for power generation, will have lifetimes shorter than the average coal mine. And when they are shut down, their magnificent gorges will be as useless for creating energy as an empty oil well' (ibid: 226).

Various solutions to the siltation problem have been tried or considered. One is to build a new dam to create a new reservoir in place of the old silted-up one. But, as Khalid Mahmood, a Pakistani expert, writes, 'in many basins, additional sites (for reservoirs) are hard to find, and in general, remaining sites for storage reservoirs are more difficult and, hence, more expensive to develop' (quoted in ibid: 230). Moreover, in general, they flood more land and produce less power than the ones they replace.

Other solutions are flushing the reservoir with river-water, dredging, building small dams upstream to catch the sediment before it reaches the main dam, and blowing up the old dam with dynamite and building a new one in the same place. All these methods are very costly. According to an expert, debris dams can cost more than the main dam.

Dredging costs roughly 20 times the cost of building a reservoir of equivalent capacity (ibid: 232).

Soil conservation measures, such as tree planting, may be effective in reducing siltation if the catchment area is small, but not if it is large. Mahmood writes: 'High concentrations of sediment (in reservoirs) are largely associated with climatic, tectonic and geological factors', and 'Sediment yields are largely unaffected by watershed management' (ibid: 231). So one should not think in a simplistic manner that as long as the rivers flow, dams will provide us with power. Pearce concludes: 'The truth is that if hydroelectricity is to be genuinely renewable, it will rarely be cheap. If it is to be cheap, it is not renewable' (ibid: 232).

Wind energy

Wind energy proved its viability in the pre-industrial age. But generating electricity from wind is more difficult. Firstly, a wind turbine is a high-tech product, so it costs a lot of money. Secondly, except in very windy regions, wind-electricity cannot compete (yet) with electricity generated from fossil fuels. Thirdly, the process of transforming the kinetic energy of wind into electricity involves wear; the average lifespan of a wind turbine is only 10 years (Melíß *et al.* 1995: 59). Fourthly, the wind does not blow all the time, nor when we want it to. This irregularity means that nuclear or fossil-fuel power plants must always be kept ready to guarantee the electricity supply when the wind is not blowing. Some critics say also that it is not possible to shut them down temporarily or to lower their production when the wind is blowing. This means that there is a lot of wastage and that wind turbines cannot replace any conventional power plant, nor cause a reduction in CO_2 production (Sünkens 1995: 113). (This also applies to solar electricity on cloudy days.) Such an objection can be invalidated if wind farms are packaged with gas turbine or hydro-power plants, but this, of course, involves double capital investment. Wind-electricity needs subsidies to be marketable. In Germany, a law compels the electricity companies to take any electricity that wind turbines generate at DM0.17 per kWh (*Frankfurter Rundschau*, 12 October 1995). In addition, there are investment subsidies.

In the USA in the early 1990s, at an average wind speed of 6 metres per second, the generating cost was 7–9 cents per kWh compared with 4–6 cents for coal- or gas-generated electricity (Flavin 1995: 60). So in respect of the cost discrepancy, wind-electricity is in a much better position than solar electricity. In Germany, in areas where wind conditions are very good, especially in the coastal areas, electricity generation through wind turbines has become a profitable business, as the cost of production there is much lower than the guaranteed price. In such areas, people would set up wind turbines even without investment subsidies.

I have not found any data on the energy balance of wind-electricity. But the facts presented above indicate that the energy balance is positive at suitable sites. But wind-electricity may never become as cheap as electricity from fossil fuels, because the supply is intermittent and hence stand-by generators are required. Stand-by generators are also required for thermal and nuclear power. But such power supply is not normally intermittent.

RECYCLING

Since in future energy will not be cheap and abundant, the dream of winning reasonably priced resources from ordinary rock and/or low grade ore must be given up. But recycling has long been known to be useful. Energy cannot be recycled directly. But through recycling materials, the energy contained in them can be reused. One must not think, however, that it is only a matter of changing our habits. In the ecology movement there are many illusions in this regard. In the 1970s, André Gorz hoped that 'entire' amounts of raw materials could be recycled or reused (Gorz 1983: 79). In the ecology movement one says: waste in dumps is a resource in the wrong place. But there are limits to recycling. A report to the Club of Rome quotes from a 'Report of a NATO Science Committee Study Group, 1976', entitled 'Rational Use of Potentially Scarce Metals':

> Whereas many kinds of uses ... leave the metals in such concentrated form that ... nothing stands in the way of their reuse, in case of a large number of metals, they are used in such a way (e.g. zinc as additive in paints) that any kind of recycling is out of the question (dissipative use). Between these two extremes, there are numerous uses in which recycling is, of course, not impossible but quite difficult, so that at present, mostly for economic reasons, one does not recycle them. ... of these metals, which are becoming ever scarcer, still on average about 70 per cent of the annual production gets lost. Even if the remaining 30 per cent is recycled ... after 10 'life cycles' only 0.1 per cent thereof remains usable. (Gabor *et al.* 1976: 144–5)

Recycling is difficult for two reasons. Firstly, dissipation is the unavoidable result of what Georgescu-Roegen called the 'fourth law of thermodynamics', which states that in the economic process, the entropy law applies also to matter (cf. Georgescu-Roegen 1986). That is, materials which are concentrated at the beginning of an economic process get dissipated and mixed up with other materials. Separating and collecting them again for the purpose of recycling requires the expenditure of energy, labour and materials. All of that – and this is the second reason – costs money. Since in future energy will become dearer, recycling will become more difficult.

Much can be done through increased use of human (and animal) labour. If human labour is not too costly, the rate of recycling can be quite high, which is the case in most third world countries. In countries like Germany, too, the rate of recycling of used paper and glass is relatively high, because many citizens bring these things free of charge to the container.

The recycling rate of any material depends on several factors: its price, the price of substitutes, energy, and equipment, the prevailing wage rates, technological development, and so on. The data given by Gabor *et al.* are from the mid-1970s. But even in the early 1990s, the recycling rate of aluminium in Germany was almost zero, of plastics only 8.6 per cent, of used paper (related to paper consumption) not more than 37 per cent, and of used glass (related to the sale of glass receptacles) not more than 30.6 per cent (Michelsen & Öko-Institut 1991: 286). Unless technological developments assume miraculous proportions, making recycling much cheaper than today, the situation will not improve as long as the matter is left to market forces in capitalism.

But recycling is necessary, not only because resources are becoming ever scarcer but also because it contributes towards protecting the environment by reducing the quantity of materials that have to be processed to produce anything. Before we take up this point, let us examine technological environmental protection, which has been the dominant policy for the purpose.

TECHNOLOGICAL ENVIRONMENTAL PROTECTION

Of course, investment in relevant technological progress – that is, research and increased use of known technologies – should enable us to reduce pollution directly. However, such a policy generally does not *solve* the problems but only achieves their medial, local, and temporal displacement. Pollutants are intercepted by means of technological processes before they escape into the atmosphere, and then dumped on the ground. Or, conversely, waste is burnt and pollutants are released into the atmosphere. Or, to prevent pollution of rivers or land sites, pollutants are dumped in deep sea. Through high chimneys, pollutants can be thinly dispersed over a large area. They can be diluted by adding fresh water or air. And they can be dumped in poor third world countries. Helmut Weidner concludes:

> an environmental protection policy that is not oriented towards the overall ecological context ... can, of course, by means of selective and peripheral interventions, achieve short- and middle-term improvements, but in the long run, the achieved successes may prove to be ephemeral or the old problems reappear at a higher level. The explanation of this ... lies in the accumulation of residual pollutants generated by the process of economic growth and, above all, in the phenomenon of problem-shifting. (Weidner 1985: 184)

In 1976, the then President of the West German Federal Environment Office, von Lersner, referred to his 'pure Sisyphean work'. He said: 'By the time we have brought one pollutant under control, another has become a problem' (quoted in *Der Spiegel*, No. 40/1976: 62).

Nowadays, rich countries dump much of their toxic waste, or even ordinary waste, in third world or East European countries, or they transfer environmentally problematic industries to such countries. So the rich countries may have fewer problems today than in 1976, but seen globally, technological environmental protection cannot be called a success.

In the context of technological environmental protection, reduction in pollution at particular points of production is generally achieved by using filters and other equipment, which are all industrial products. Their manufacture and operation require, as with all industrial products, considerable expenditure of energy and other materials. That also causes pollution, but somewhere else and of another kind. For example, sulphur dioxide emission from a power plant can be largely eliminated, but it requires a chemical plant that consumes 3 per cent of the power produced (Hayes 1979: 6; Trainer 1985: 108). This means that, *inter alia*, more coal has to be burnt, which leads to more CO_2 emission. Moreover, filters and similar items have a limited life, and must be replaced every 10, 15 or 20 years. Protection of the environment thus becomes a regular industry, causing more resource consumption and more pollution elsewhere.

It is conceivable that in the initial phase of a technology, pollution per unit of production can be reduced through further development of the technology – through new ideas and without more filters. But at some point the optimum will be reached, the technology will attain maturity. After that, increase in production will be accompanied by at least a proportional increase in pollution, which would necessitate the use of more filters, etc.

Of course, it might be possible through chemical methods to transform many pollutants into harmless substances. But, firstly, it is impossible to intercept all kinds, or entire quantities, of pollutants, and secondly, such transformations require, *inter alia*, about the same amount of energy as the preceding production process (Meyer-Abich 1973: 177). This makes the idea both too costly and ecologically meaningless, as more energy production causes more pollution, particularly emission of more CO_2, and more radioactive particles.

'ECOLOGICAL' INDUSTRIALISM

In recent years, many ideologues of industrial society have realised that limits to resources and limits to the capacity of nature to neutralise pollution are real. They have also realised the limits to technological environmental protection. Nevertheless they continue to believe that industrial economies need only to be adjusted to new constraints. For Hardin B.C. Tibbs, the aim of what he calls 'industrial ecology' is

> to interpret and adapt an understanding of the natural system and apply it to the design of the manmade system, in order to achieve a pattern of industrialization that is not only more efficient, but that is intrinsically adjusted to the tolerances and characteristics of the natural system. The emphasis is on forms of technology that work *with* natural systems, not *against* them. ... Our challenge now is to engineer industrial infrastructures that are good ecological citizens, so that the scale of industrial activity can continue to increase – to meet international demand without running into environmental constraints ... (Tibbs 1992: 6; emphasis in original)

In the following, I shall present and discuss selected aspects of this new ideology.

Industrial clustering

In what is actually only a further development of the idea of recycling, Fritjof Capra, like many others, calls upon human beings to imitate the organisational pattern of ecosystems:

> We must build up our industries in such a way that the waste of one industry becomes raw material for another. Wastes and resources in the whole system must be passed on cyclically. ... A brewery, e.g., generates organic residual materials which serve as nutriment for mushrooms and worms. The worms are used for feeding chickens. Mushroom culture finally generates waste that serves as cattle fodder. That is a genuine cyclical flow in industry. (Capra 1996)

Capra also gives the example of small steel works in the USA which use the scrap iron of nearby cities as raw material (ibid.).

In Kalundborg, Denmark, a thermal power plant is supplying process steam to an oil refinery and a pharmaceutical plant, surplus heat to households in the district and to its own fish farm to warm the sea water, fly ash to cement producing and roadbuilding companies, and it is desulphurising its smoke using a process that yields calcium sulphate as a by-product, which it sells to a wallboard factory. The oil refinery supplies cooling water and its purified waste water to the thermal power plant, its surplus gas – after removing the excess sulphur – to the wallboard factory and to the thermal power plant for burning. The sulphur is supplied to a sulphuric acid plant. Sludge from the fish farm is going to farmers for fertiliser. The pharmaceutical company is treating its sludge chemically and heating it at 90°C for one hour to produce liquid fertiliser free from micro-organisms, which is being supplied to farmers (Tibbs 1992: 9).

This 'multidimensional' recycling must be supported. But it is just recycling, not a miracle. No third world person would find anything new in small steel plants using scrap iron from nearby cities as raw material. Any old European who lived in his youth in rural areas knows that potato peelings from kitchens used to be given to pigs as fodder, and that scrap dealers used to come to every village to collect all kinds of scrap. The question that ought to be asked is why this practice disappeared from Europe and why it continues in India. The answer is simple. Collecting, separating and then processing scrap is labour-intensive. If wages are high, it may not be economical to collect scrap from widely dispersed sources. In India, wages are sufficiently low.

In Kalundborg, the transactions are based strictly on price. They are stimulated by mandatory cleanliness levels and fees for hazardous waste disposal (ibid: 9–10). That is fine, but no surprise. Capitalists look for economic advantage everywhere. There is, therefore, no reason to call such transactions by as high-sounding a name as 'industrial ecosystem', as Tibbs does. Nor can they be described as 'cyclical flow' or 'closing loops', as Capra, Tibbs, and the Friends of the Earth Netherlands do (ibid: 8, 9; Brakel & Zagema 1994: 15). The energy supplied by the power plant in Kalundborg cannot come back to it. It is used only a second time before it, following the entropy law, dissipates irretrievably. And the carbon that has been emitted into the atmosphere in the form of CO_2 cannot at any reasonable cost be retrieved to make coal or oil again. The same is true of the fish farm's fertiliser-sludge.

Dematerialisation – the question of materials balance

In the meantime, even among ecologists who are no critics of industrial society and capitalism, there is a consensus that, generally speaking, an economy's impact on the environment is roughly proportional to the amount of resources it consumes. This can be illustrated by the fact that in the USA, in the normal course of the economic process, that is, even without any big accident, every year 42 billion litres of oil spill out through leaks in pumping installations, pipelines and tanks. This conclusion of a study made by Friends of the Earth has, of course, been rejected by the oil industry, but it has been confirmed by the government's Environmental Protection Agency (EPA) (*Frankfurter Rundschau*, 18 May 1993). It is obvious that if oil consumption in the economy goes down, pollution through oil spills will also go down. Ultimately, therefore, the general policy for environmental protection should, according to these ecologists, be to increase resource productivity and to reduce resource consumption; this is also called dematerialisation.

Friedrich Schmidt-Bleek, of the renowned Wuppertal Institute for Climate, Environment, and Energy, has suggested 'a universal measure for assessing and comparing the environmental impact of infrastructures, products and services', namely 'material intensity per unit of service' (MIPS). MIPS takes into account the total materials and energy input in kg or tons for the whole life-cycle of the object concerned, including all the plant, transportation and packaging, from the first use of a caterpillar onwards, necessary for the production, use, and final disposal of the object. Total service units of, for example, a car are the total person-kilometres that can be obtained during a car's life-span. The German average would be 180,000 person-kilometres (150,000 km x 1.2 persons) (Schmidt-Bleek 1994: 98).

Energy measured in kg or tons seems strange. But energy comes from oil, coal, wood, uranium – from materials (the exception is sunshine) the weights of which can be measured. That is why we can also speak here simply of materials input. This refers not only to materials actually used in the production of a service unit, but also, for example in mining, to the overburden moved in order to get at the valuable mineral, and waste that is disposed of in dumps.

Schmidt-Bleek knows that MIPS cannot satisfactorily express all kinds of environmental impact. So he does not exclude the possibility of complements, such as a measure for land use or a measure for eco-toxicological impact. With this approach, in my opinion, Schmidt-Bleek accepts Georgescu-Roegen's methodological demand for calculating the energy balance of any energy-producing technology, and goes further. However, there is a difference between energy balance and materials balance. In the case of the former, one can say whether the balance is positive or negative, because the product on both the input and output side is the same – energy – sometimes even the same form of it – electricity. But in the case of materials balance, on the input side is, say, iron ore, and on the output side a car – qualitatively very different things. But one can compare the MIPS of a car with that of a bullock cart. And that is also very important.

Recycling as a method of reducing MIPS has already been discussed. But modern defenders of industrial society lay more emphasis on technological progress. They speak of an 'efficiency revolution', which could enable us to derive more prosperity from less materials flow than we normally do today. This was also the main hope of the authors of the Brundtland Report. In the mid-1980s, they noted 'some favourable trends' that allegedly proved that 'future patterns of agricultural and forestry development, energy use, industrialisation, and human settlements can be made far less material-intensive, and hence both more economically and environmentally efficient' (WCED 1987: 89–90). Chapter 8 of the report carries the title 'Industry. Producing More with Less'.

In their recent report to the Club of Rome, Ernst Ulrich von Weizsäcker, Amory Lovins, and Hunter Lovins give numerous examples of successfully reducing MIPS by a factor of four (Weizsäcker *et al.* 1997: part one). Their claim should be noted that in no case has there been a loss in comfort, convenience, or quality of the service concerned. On the contrary, in most cases, there has allegedly been an increase in them. What is more, it is claimed that everything costs less than before, and those who implemented the innovations, involving mostly high technology, even profited financially.

Any method of reducing MIPS should in principle be welcomed. But, especially in the case of high-tech methods, we should take care that euphoria does not blind us to the fact that MIPS at one much-publicised point can be reduced at the cost of an increase in MIPS at other, unnoticed points. For instance, energy consumption can be reduced by increasing materials consumption for insulating houses. Prices and financial balances, as explained earlier in this chapter, are not always relevant indicators in this matter, because they depend on so many other factors. Of course, MIPS has been conceived in such a rigorous manner that the above warning should not be necessary. But energy balance is also a rigorous concept. Yet we have seen in the case of solar electricity how widely divergent and doubtful calculations of energy pay-back time can be in practice. Moreover, a company may be able to reduce MIPS at particular strategic points of its operations and so become more efficient. But with this increased efficiency, production and sales may register additional growth, so that the net effect on the environment may be zero or even negative.

Arriving at figures of MIPS and MIPS reduction is very difficult, even for scientists. So I shall not try to examine the claims contained in the Weizsäcker report. But a few doubts about the quality of the claims made in connection with the various examples should be expressed:

(a) Weizsäcker *et al.* give the examples of two aircraft manufacturers – Boeing and Lockheed – who spent a lot of money to raise the efficiency of the lighting system in the engineers' offices and on the shop floors. They happily report that, in the case of Boeing, workers could thereafter see the defects and mistakes in the construction process much better, so that delays in production could be avoided. At Lockheed, as a result, absenteeism went down, labour productivity increased, and running costs could be reduced, so that the company could win a big contract in the face of tough competition (Weizsäcker *et al.* 1995: 76–7). This is very good for the companies, but what did the environment gain? The improvements contributed towards growth in aircraft production, and so to growth in materials flow.

(b) The authors give the example of a Dutch bank: International Netherlands Group (ING), formerly Nederlandsche Middenstandsbank (NMB). In the last 15 years, the bank achieved a high growth rate which, according to the authors, had much to do with resource efficiency. The management got a new head office building, which combined natural materials, plants, sunshine, art, quiet, and water in an organic way. It was made 10 times as energy-efficient as the old building. So the bank got a new image. The volume of business grew dramatically, and it became the second largest bank in Holland (Weizsäcker *et al.* 1997: 27–9). This is very good for the bank; but is it also good for the environment if the volume of a bank's business grows? Are we supposed to believe that the Netherlands' second largest bank finances no business that destroys nature and wastes resources?

(c) The authors give the example of Davis, the US town famous for its eco-consciousness and its Village Homes Project, with 200 houses on 28 hectares of land. There are many praiseworthy ecological features in this 'village' in a town: narrow streets, parks with fruit trees, biological sewerage, and so on. The residents cover most distances on foot or bicycle. But the villagers 'naturally' use solar electricity, which has a negative energy balance. The houses have two entrances – one at the front for pedestrians and the other at the back for cars, one of the worst enemies of the environment. It has not been possible to attract normal job opportunities (ibid: 133–4). Why? Because, naturally, the 'villagers' want to earn the high average US income, but jobs that can offer such incomes generally come from industry and business that does not think small is beautiful. The 'villagers' drive to their jobs, instead of living where the jobs are.

The greatest economic gain is that the village's house prices register the highest rate of growth in the town. But what is so very special about it? Houses in greenery with less traffic than usual have always attracted rich people, who often want to possess more than one house – one in the big city, the other amid greenery. Where is the reduction in total materials flow?

(d) I also have doubts about the ways the authors have calculated. For instance, they speak of a car combining an internal combustion engine with an electro-motor. The latter can collect 70 per cent of the 'braking energy' and supply it as additional power for uphill drives or for acceleration (ibid: 8). This way, of course, energy efficiency can be increased. But two motors in a car? Will that reduce MIPS?

To take another example, the authors speak of a house in England, built using 'traditional labour-intensive' methods. Construction cost per square metre was, 'compared to current prices, low' (ibid: 24–5). If that is true, I wonder why construction companies in Europe lay off workers and use machines to save wage costs.

(e) The authors give an example of bio-intensive small-scale agriculture with traditional techniques. That is all very well. After all, the destructive aspects of modern large-scale agriculture are now well known. The authors say the farmer Jeavons and his colleagues need only 180–360 square metres of land to produce the total food needs of a vegetarian person. They need no machines, no chemical fertilisers or pesticides, and much less water than in the usual kind of agriculture. That is a lot of dematerialisation. The labour requirement is also surprisingly low. Since only relatively bad-quality land and a few agricultural tools had to be bought, the per-hectare income is double that of the usual kind of agriculture. The authors themselves admit that that is not much, for in conventional agriculture, calculations are made on the basis of the wages of full-time workers (ibid: 99–101).

The authors admire some traditional bio-intensive agricultural and horticultural techniques practised in China and warm Asian countries, where the per-hectare yield is much higher. But at the same time: 'For the sake of truth, one must however admit that a conversion to the principles of Jeavons cannot take place in the rich industrialised countries without substantial changes in our present-day lifestyle' (Weizsäcker *et al.* 1995: 133). Indeed, and the agricultural workers of these countries would also have to accept Chinese wages. 'But', the authors write, 'who can say that today's agricultural economics would still be valid when in an overpopulated world with billions of unemployed people only per-hectare yield will matter?' (ibid: 132). This is correct. But then the authors cannot at the same time logically promise that individuals, firms, and countries who follow their advice would not suffer any loss in comfort and would even become richer.

So ultimately we come to the question of lifestyle. In regard not only to food and agriculture, but also to other resources, the authors have to advocate change in lifestyle. They give the example of a hotel that used to change towels and sheets every day. The management asked the guests whether they wanted this, and the great majority of them said no. The hotel changed its practice, and the annual water and power costs fell by US$70,000 (Weizsäcker *et al.* 1997: 89). The authors themselves now advocate – in addition to installing super-efficient high-tech equipment – such simple solutions as switching off lights that are not necessary, using pressure cookers, wearing warm clothes indoors, and so on; that is, a reduction in energy comforts of 3 per cent (ibid: 64).

Most of the examples given by Weizsäcker *et al.* come from the USA and rich European countries. With the notorious wastage of resources practised in these countries, it is not very difficult to achieve a reduction in resource consumption by a factor of four by means of a combination of technological improvements, a little reduction in wastage, and a little sacrifice of comforts. But Schmidt-Bleek (and others) demand a reduction by a factor of ten in western industrialised countries, so that the underprivileged 80 per cent of humanity might enjoy a somewhat higher standard of living, and so that global materials flow may be halved (Schmidt-Bleek 1994: 103). Wolfram Ziegler thinks that the same degree of reduction would be necessary just to stop the exponential growth in the rate of extinction of species (quoted in Bahro 1987: 30). Will such reduction be possible through technological efficiency and a little reduction in wastage?

POSTINDUSTRIALISM, THE INFORMATION SOCIETY, AND THE COMPUTER

In connection with dematerialisation, one hears so much of the 'information society', 'postindustrialism', 'post-maturity economy', 'service society', etc. One points out that, for example, in the USA more than 60 per cent of employed people only process information in some form or other (Naisbitt 1982: 14). Referring to this fact, some even speak of dematerialisation in advanced economies as if it has already largely taken place. That this is utter nonsense was realised by the authors of the Brundtland Report: '... even the most industrially advanced economies still depend on a continued supply of basic manufactured goods. Whether made domestically or imported, their production will continue to require large amounts of raw materials and energy ...' (WCED 1987: 217). Moreover, neither processing information nor selling services is possible without the use of material goods.

Here, common sense is useful. Firstly, if energy- and materials-intensive branches of industry are transferred to underdeveloped countries, then of course the energy and materials balance of the highly developed economies looks better, but that of the world economy remains unchanged. (The same applies to environmental degradation.) It is a zero–sum game. Secondly, if one examines not only production but also consumption, then it becomes clear how bad the balance of the highly developed economies is. For one unit of consumption, say quenching thirst, an information-processing US citizen uses up many times more resources, such as cola in a can, than an average citizen of India, who drinks a glass of tap water. The data-processing US citizen consumes cars, meat, refrigerators, holidays on a sea-beach, and so on. One cannot eat data, nor live in a computer.

Even if they concede the above critique, the ideologues of ecological industrialism and large sections of the ecology movement believe that the computer and, more generally speaking, information and communication (IaC) technology not only themselves require much less resource input compared to earlier technologies giving comparable services, but also contribute massively towards reducing resource consumption in many other branches of the economy. They are also believed to be environmentally clean technologies. Is this true?

The computer, perhaps the most important element in IaC technology, is small, and its use requires very little energy. But this is not decisive. A study carried out by Jürgen Malley of the Wuppertal Institute shows that very large amounts of energy and materials are necessary to produce a computer. It requires a large number of pure and partly precious materials, for the production of which large amounts of energy and materials are necessary. Taking everything together – production, use for four years (its average life-span)[3] and recycling (if undertaken) – a personal computer necessitates in its life-cycle, from the first use of a caterpillar to final disposal, the use of 15–19 tons of materials, depending on the kind of use. In addition, just to produce its various parts, 33,000 litres of water are required. Producing a car requires only a little more material (25 tons) (Malley 1996: 48).

The figure of 15–19 tons is a total. To get a computer's MIPS, this figure must be divided by the average total service the computer gives. But that is very difficult to calculate, for, unlike a car or a refrigerator, a computer gives many different kinds of service. In any case, the fact that a computer, with an average life-span of only four years, needs so much material input should refute the notion that it is a resource-efficient apparatus. What should one say about the resource-efficiency of a communication satellite, another very important element in IaC technology? Who knows the total resource input required by a PV cell, a very important element in a satellite?

Of course, the more intensively computers and satellites are used, the lower their MIPS – as in the case of any machine. For instance, if a jumbo jet transports 400 holidaymakers, its MIPS is lower than if it transports only 200. But at other points, 400 holidaymakers consume more resources than 200 holidaymakers. An ecologist's main concern should, therefore, be the total materials flow in the (world) economy as a whole, because it is mainly that which determines the economy's total impact on the environment and the latter's general health. Does IaC technology contribute towards reducing the total materials flow?

The answer is clearly no, not if other things remain the same. Generally speaking, faster and more efficient data processing through computers leads not only to more efficient, but, other things being equal, also to accelerated and increased flow of materials. With the help of IaC technology, large companies can co-ordinate their production and sales at various places in the world. This furthers globalisation of production and the market, and acceleration of sales. More and more new products can be developed faster and faster. Hitherto unexploited labour can be exploited, labour productivity can go up, and, although resource depletion goes on, the price of many products may go down. Moreover, computer technology makes constant fashion changes and improvement of old products easier. New generations of products very soon replace similar older products, so that the latter are discarded prematurely. The life-span of products becomes ever shorter. All this leads to more consumption and larger materials flow.

If IaC technology could really make some economic activities superflous, then it might perhaps contribute towards reducing the total materials flow. But, according to Rainer Grießhammer of the Eco-Institute of Freiburg, 'Only a few products and uses are substituted through telecommunication. It mainly leads to new products and uses and so inevitably to additional environmental impact. New technologies mostly have additive effects' (Grießhammer 1996: 53). This is true. The TV has not replaced the radio, nor has the computer replaced books and paper. A large part of the growth in demand for office paper has been caused by desktop computers and photocopiers. The 'paperless office has failed to arrive' (Tibbs 1992: 14).

Much has been expected from so-called telematic. A combination of telecommunication and information helps the traveller to make an optimal choice of the means and paths of travel. He/she gets all the information on public transport, but also on traffic jams and diversions, so that motor traffic flows without hindrance. But, writes Heinz Klewe, area planner and traffic expert, 'Why should a potential road user whose car stands in front of the house decide in favour of bus or train when the on-board computer promises to guide him/her individually around construction sites, traffic jams, and accidents?' (Klewe 1996: 36). In fact, by means of telematic, more road traffic can be optimally handled, which generates more traffic.

Much reduction in traffic and in materials flow has been expected from tele-homework. But, firstly, the potential of such work has been overestimated. At the moment, in Germany, there are only about 3,000 such jobs (cf. Schütte 1996: 27). Secondly, most tele-workers have two places of work – one at home and the other in the office. Of course, he/she does not drive to the office every day, but office space, furnishing and a computer are needed in two places. This means double materials expenditure. Moreover, 'one who has to drive to the office only once a week, tends to accept a longer distance' (ibid: 28). Further sub-urbanisation and further motorisation, both spoiling the landscape, are unavoidable consequences.

The fact that computers and most electronic goods are becoming smaller and smaller also constitutes a disadvantage for the environment. Such products are highly complex and contain a mix of several materials. Miniaturisation makes it more and more difficult, sometimes impossible, to separate these materials, and this is an obstacle to recycling. Germany generates 1.5 million tons of electronic scrap every year, 120,000 tons of which is computer scrap (cf. Malley 1996: 49). And it all contains highly toxic substances.

LAND, WATER, AND POPULATION

Land and water are essential resources. Water is renewable. But we cannot increase the amount of rain and snowfall, the ultimate source of all naturally fresh water. The area of land is given, and is inexhaustible, but its fertility can become exhausted. Of course, sea water can be converted into fresh water, but that is a very energy-intensive and hence costly process. Moreover, since energy prices will rise in future, this process would also become ever costlier. So far as irrigation is concerned, as already mentioned in connection with hydroelectricity, suitable sites for dams are scarce. As far as land is concerned, soil erosion is caused not only by human activity. It is also a natural process. Preventing erosion and other kinds of soil degradation requires a lot of energy. The same applies to land reclamation.

Given these facts, it is not easy to increase food production. In eco-alternative discourse, one often hears the following quotation from Mahatma Gandhi: 'Earth provides enough to satisfy every man's needs but not for every man's greed' (Gandhi 1997: 306). This rhetoric is too simple. It ignores the question of the number of people whose needs must be satisfied, which cannot be ignored. For while the global human population is continuously growing, the rate of growth in food production is falling. Lester Brown of the Worldwatch Institute shows that while between 1950 and 1984 world grain production grew – thanks to chemical fertilisers, pesticides, irrigation, and high-yielding varieties of plants – at the rate of 3 per cent per annum, between 1984 and 1993 the annual growth fell to less than 1 per cent, reducing per capita availability by 11 per cent (Brown 1994: 179). Brown also cites FAO experts who state that both fish catch in seas and meat production can hardly be increased any more, so that per capita availability of animal protein is tending to fall (ibid).

In view of these facts, it should be obvious that we should try to stop population growth rather than grow more food. For both population growth and growth of food production have an adverse environmental impact. Yet several political forces, with which I sympathise in other matters, oppose any active population control policy: leftists, feminists, third world solidarity groups, and third world NGOs. For reasons of space, I shall give only one example. Ursula Pattberg, a German third world sympathiser, wrote in 1992:

> I am against any population policy because it is not needed at the moment. What is needed is rather a different policy on resources. It is a matter of … distribution of land, air, water, food and other things. If we can achieve a just distribution of these resources, then the debate on population policy is superfluous.

On the ecological crisis, she wrote: '… the causes of environmental destruction in Thailand … are quite clearly the interests of capital. It is a matter of profit maximisation and destruction of nature for the sake of short-term profits.' Then she reported: 'In Thailand, there is a People's Forum in which several NGOs have joined up. In their programme, the word "birth planning" does not occur at all; overpopulation is for them no subject for discussion.' She quoted from the programme of People's Forum:

> The economically powerful countries will intensify their efforts to control the resources. They will use all ways and means including international organisations like the World Bank in order to defend their good starting position and to represent their interests. The countries of the Third world have been instrumentalised in this process – ignoring the fact that they have a right to their own development.

Pattberg sums up: 'What these groups demand is the right to use their own resources' (Pattberg 1992: 5–6).

This opposition is the result of several confusions, which must be removed.

Malthus – the difference between problem and policy

We must begin with Malthus. We must differentiate between problem and policy. Population policy can be debated, formulated, and accepted or rejected. But the population problem is an objective state of affairs, which cannot be conjured away. Any good ecologist knows that in nature, the population of any species must remain within a certain limit if it is not to upset the delicate ecological balance. Mixing up problem and policy in discussions only creates confusion. The indignation against Malthus is justified. According to him, the poor are themselves to be blamed for their poverty. But the question is whether, for this reason, Malthus' presentation of the problem is also wrong.

Malthus' harshest critics have always come from the left. Marx considered Malthus' essay to be a 'libel on the human race'. Engels wrote in 1865: 'economic laws are not eternal laws of nature but historic laws which arise and disappear'. He thought 'what is tenable in the so-called Malthusian theory' is valid only for societies 'based on class rule and class exploitation'. That was no proper refutation. But serious efforts were also made to refute one of Malthus' two laws. Engels and Lenin rightly thought that his law regarding food production, namely that it increases only in arithmetical progression, is actually based on the law of diminishing returns. But they declared that the limitless advance of science and technology nullifies the law of diminishing returns, which is otherwise valid. (This summary of the views of Marx, Engels and Lenin is based on quotations contained in Meek 1971.)

But science and technology have in the meantime disappointed expectations. Brown shows that between 1950 and 1984 the application of each additional ton of fertiliser boosted grain output by 9 tons. But this ratio started to worsen in 1984. By 1989, it was down to 1.8 tons more grain to each additional ton of fertiliser (Brown 1994: 184–5). The plants are simply not responding to more and more fertiliser. The law of diminishing returns is, therefore, not nullified.

Malthus' other law – that population grows, *if not controlled*, in geometrical progression – is more difficult to refute. The only thing that his opponents can do in this respect is to point to the fact that in industrial societies population has stopped growing or is growing very slowly. But that is also no real refutation, as most couples in these countries are using various means to control births. An average healthy couple in an industrial society *could*, if they used no contraception, produce ten or even more children in 30 years. Clearly, Malthus' two laws are as good as natural laws. At the least, they are still valid.

Scientific and technological research has not come to an end. So one may still hope that, say, biotechnology might help to solve the problem, or that there might be a breakthrough in the process of photosynthesis, enabling plants to convert solar energy more efficiently into edible energy. Gail Omvedt and Govind Kelkar, two feminists from India, report that S.A. Dabholkar, a farmer–experimenter from western India, has developed methods through which, with the aid of 'low external inputs', a family of five could enjoy a middle-class lifestyle by cultivating just one-quarter of an acre of land (1 acre = 0.4046 ha), so that 'there is at the technological level no real population problem.' They write: 'The fact is that the ability of sunlight and soil (which itself can be regenerated and even "made" from biowaste material) to achieve high productivities has hardly even begun to be tapped. ... the "carrying capacity" of the various regions of the earth can be almost *indefinitely* extended' (Omvedt & Kelkar 1995: 30; emphasis added). But Dabholkar's claim and Omvedt's and Kelkar's faith in technology sound like faith in a miracle. The authors also report that in India, Dabholkar's claim is 'generally taken as exaggerated and even fantastic' (ibid.). And Brown reports that leading scientists hold out little hope for far-reaching breakthroughs in these areas in the foreseeable future (Brown 1994: 186–7).

How acute is the crisis? How much latitude do we still have?

Let us examine some other well-known arguments against the need for an active population control policy. 'Development is the best contraceptive', has been the slogan of many leftists since the world population conference in Bucharest in 1974. This argument is also supported by the theory of demographic transition, which states – on the basis of the demographic history of the industrialised European countries – that in a society with a growing population, the birth rate falls with growing prosperity, quasi-automatically, and finally equals the death rate, thus concluding the transition to a stable population. But this theory was formulated before humanity became aware of the limits to growth. Most countries of the third world would never be able to reach the prosperity level of, say, West Germany in 1972, when that country completed the final phase of demographic transition. Moreover, prosperity alone may not suffice. In Saudi Arabia – for the last two decades one of the richest countries of the world – the birth rate was in 1989 more than 40 per thousand per year (cf. Meadows *et al.* 1992: 32).

As a kind of supporting argument to the above, leftists argue that the problem of hunger in the present-day world is no problem at all, that there is enough food, which needs only to be distributed properly. That may be theoretically perfectly true. However, some questions remain:

- We are concerned not only with the situation today. We must also ask: how long can world food production keep pace with a growing world population? The facts and figures quoted from Brown give no grounds for optimism in this regard. Can we, like believers in a religion, simplistically hope that science and technology will somehow solve all future problems?
- The negative ecological effects of intensive agriculture are well known. Is it permissible to intensify it still further in order to produce more food for a growing population?
- Can we expect the peoples of the food-surplus countries to work hard and invest their money in order finally to give away their surplus to the poor of the third world? The world has not become one communist world yet!
- The farmers of the food-surplus countries would certainly like to sell their surplus. But from where will the poor countries of the third world get the foreign exchange to pay for the continuously rising food imports?

An alternative line of argument used against any active population control policy is based on calculations pertaining to the maximum carrying capacity of the Earth. In 1982, an FAO and UNFPA study asserted that there is enough land in the third world (without China) to feed 33 billion people – but only if every square metre of cultivable land and large quantities of fertilisers and other chemicals are used to produce just a sufficient quantity of vegetarian food (cf. Sadik 1990: 7). There is also a model for the production of sufficient food for 15 billion people with moderate use of chemicals. This model, it is asserted, permits an ecologically careful handling of nature (cf. Simon 1991: 30) It is generally assumed that the world population will stabilise at some point between 2050 and 2100, at 11–14 billion. According to these models, therefore, there is not only enough time and latitude for the demographic transition but also no reason at all for panic.

I have some objections to this line of argument:

(a) If we in the third world (without China) want to produce enough food for 15 billion people with just a moderate use of fertilisers and other chemicals (because we do not want to damage the environment), then agriculture must become extensive. We would need more cultivated land, and also more land for houses, roads, schools, offices, factories, and so on. But, writes Brown, 'the reality is that there is simply not much fertile land waiting to be plowed' (Brown 1994: 187).

There is a difference between land and fertile land. In the wake of the doubling of world grain prices in 1972, the farmers of the world increased the area under grains by 11 per cent. But this was followed by a massive retrenchment. A very large part of this increase took place on land highly liable to erosion and incapable of sustaining cultivation. As a result, in the territory of the former Soviet Union, the grain-harvested area declined from 123 million hectares in 1977 to 99 million ha in 1993. In the USA, between 1985 and 1992, 14 million ha were again converted to grass or trees (ibid: 182).

Brown does not give comparable figures from the third world. But we know that, in Brazil, agriculture on land previously covered by rain forest is possible for only 3–4 years. Such land is fast eroded. There may still be a lot of such land in the world that can be brought under the plough. But, according to one estimate made in 1991, 6 million ha of cultivated land is being lost every year through erosion, salination and so on (Stiftung Entwicklung und Frieden 1991: 237–8). According to Friends of the Earth Netherlands, 'current production methods will lead to a decrease in the amount of potentially arable land by about 16 million ha per year' (Brakel & Zagema 1994: 16). As far as the third world is concerned, Sandra Postel writes that since 1945, in Asia, Africa and South America, land degradation induced by humans amounted to 20, 22, and 14 per cent respectively of the total vegetated land (Postel 1994: 10).

Let us suppose that by means of great effort and a proper mix of correct policies, this trend can be stopped. Even then, feeding 15 billion people by means of extensive agriculture would mean that a large portion of the remaining forests would be lost. Of course, the demands of luxury industries could be rejected. But even satisfying the other basic needs of 15 billion people – firewood, building timber and paper – would lead to the continuous destruction of ever more forest. Moreover, apart from the fact that we humans need there to be a certain proportion of land covered with forests, these forests are the habitats of many other species (which we also need). Does the human species have a right to conquer more living space?

(b) In that case, millions of people would have to migrate into areas that are still thinly populated. But these areas belong to other groups of humans – to the Amerindians, Maoris, Aborigines and other tribes. Should they be pushed out of their land again? Should we wage war against them?

(c) What should the people of the already densely populated countries – such as India, Bangladesh, Egypt – do, if the peoples and rulers of the thinly populated countries do not give them the permission to immigrate? Of what use then are the FAO and UNFPA models?

(d) Land is not the only problem. With only rain-fed agriculture, food production cannot be increased much. Irrigation would be necessary. But from the Worldwatch Institute to the World Bank, everybody knows that in many parts of the world fresh water has already become a scarce resource. There is fear that agriculture and industry as well as humans and nations would fight each other for water. There would be social conflict. The World Bank fears that in the next century, wars may break out between neighbouring countries on the question of access to fresh water (*Frankfurter Rundschau*, 31 December 1996). It is unthinkable to transport fresh water from, say, Canada to India.

(e) Finally, opponents of any population control policy hold that the preconditions to producing enough food for 15 billion people on land in the third world (excluding China), if it is at all possible, are a sound agricultural policy, egalitarian economic development, a different development strategy, and so on. These would constitute the minimum social conditions under which population growth could slow down gradually.

But it is uncertain whether fulfilment of these social conditions alone would be enough to halt population growth. Moreover, *egalitarian* economic development is not only a precondition for something but also a constituent goal of an ideal which can be attained only through a long struggle for social change. Also the other preconditions may not be soon fulfilled. If the population continues to grow in the meantime, then the ecological balance will be restored by nature itself – through hunger, war, civil war, social chaos, disease, epidemic – which nobody wants. So an effective and active population control policy is urgently necessary.

SUSTAINABILITY

Before closing this chapter, let me make some concluding remarks on sustainability. Nowadays, even many protagonists of ecological industrialism are demanding a drastic reduction in resource consumption in industrial countries. Their main purpose is to attain sustainability, but they also want to give the third world an opportunity to develop. At the same time, they are promising the citizens of the first world that they would not, as a result, suffer any substantial loss in their standard of living.

Friends of the Earth Netherlands, who start from the moral position that each country has a right to the same amount of environmental space[4] per capita and accept the consequence that consumption of natural resources in rich countries must be cut back by at least 70 per cent (Brakel & Zagema 1994: 6), write: 'The question: "can the rich North have a *reasonable* standard of living as a result of an efficiency revolution …?" can … be answered positively' (ibid: 26; emphasis added). The term 'reasonable' is subject to interpretation. A few pages later they write that 'the end result of an efficiency revolution is a *higher production and consumption* level' (ibid: 31; emphasis added). Then they proudly quote a leader of the biggest Dutch trade union, who praised their scenario because 'it is based on the solidarity principle and also guarantees a *continuing high level of consumption*' (ibid: 30; emphasis added). But then, strangely, they also demand: 'it is naturally the job of government to carry out a *sufficiency revolution*' (ibid: 32; emphasis added). Could their confusion be any worse? Weizsäcker *et al.* and the President of the Club of Rome say that prosperity can be doubled while resource consumption is halved (Weizsäcker *et al.* 1997). And Herman Daly, who for ecological and resource-related reasons advocated a steady-state economy in the 1970s (Daly 1977), wrote in 1989: '… with no reduction of the quantity of goods provided to the [US] consumer, there can be a considerable reduction of economic activity' (Daly & Cobb 1990: 296).

Ecologists cease to be ecologists when they start thinking and talking like professional politicians who must promise good things to voters. Weizsäcker writes in another book, which carries the sub-title *Ecological Realpolitik at the Threshold of the Century of the Environment*: 'To tell Europeans, Americans, and Japanese that they should wear sackcloth and ashes and forgo prosperity, is a strategy condemned to failure.' He therefore proposes a 'new kind of economy' and a 'new model of prosperity' which 'could be extended to five or ten billion people' (Weizsäcker 1989: 14). He concretises this model with an example: small 'citycars' and bicycles could be carried along with passengers in buses and double-decker trains (Weizsäcker 1994: 184–5). Friends of the Earth Netherlands try to present their scenario as 'more attractive than a "business as usual" scenario, even for business' (Brakel & Zagema 1994: 36). It must, however, be mentioned to the credit of Daly and Cobb that after describing their utopia they also write: 'Perhaps this scenario is too optimistic' (Daly & Cobb 1990: 297).

I strongly doubt that these promises have any plausible basis. The authors quoted seem to believe that the potential for technological progress is unlimited, and that, therefore, both resource productivity and labour productivity can be increased simultaneously. This is an illusion. After the optimum combination of labour-intensity and resource-intensity has been reached, the one productivity can be increased only at the cost of the other. Workers using pneumatic hammers have, of course, a higher labour productivity than workers using ordinary hammers. But pneumatic hammers surely have a lower resource productivity than ordinary hammers, for the former require very high material inputs. In any case, a model of prosperity that contains citycars in double-decker trains for five to ten billion people *and* drastic reduction in resource consumption appears to me to be Eurocentric wishful thinking and not a realisable scenario in a sustainable world economy.

It cannot be predicted what standard of living for how many people would be possible if materials flow in the industrial countries is reduced by a factor of ten or four or three. (Weizsäcker has not said how many people in his scenario would be able to afford a citycar.) But the prediction is unnecessary. We have to accept the absolute ecological necessity of this reduction, and then see how much prosperity for how many people we can derive from the remaining materials flow.

If, thanks to an efficiency revolution, prosperity can be increased and extended, well and good. If it were reduced, we would have to accept that. After all, we do not want to ruin the environment. One thing, however, is very important: we should not spread illusions. In any technology, there is an optimum point of efficiency. The laws of physics, chemistry, biology, and thermodynamics cannot be annulled. Miracles happen only in story books. We cannot imagine a car that needs no energy. And oil tankers are not becoming smaller and smaller while transporting more and more oil.

We often hear the term 'quality of life' instead of 'standard of living' or 'prosperity'. For instance, in 1991, the International Union for Conservation of Nature defined sustainable development as 'improving the quality of human life while living within the carrying capacity of the supporting ecosystem' (quoted in Irvine 1995b: 5). If we live like that, we shall automatically enjoy an improved quality of environment. A few more things – love, friendship, justice, solidarity, social peace – improve our quality of life and do not require an increase in materials consumption. But we must be careful: 'quality of life' may also be understood as material comfort, luxury, and activities that cause an increase in materials flow, such as tourism.

As far as non-renewables are concerned, even zero growth is no solution. If consumed at a lower rate, they will be available for longer, that is all. We cannot bequeath to the next generation as much of them as we inherited unless we totally refrain from consuming them. But we can consume them frugally in order to gain enough time for the transition to a sustainable economy and in order to make it as painless as possible.

Since, by definition, 'a sustainable society is one that can persist over generations' (Meadows *et al.* 1992: 209), in the *ultimate analysis*, and *theoretically*, an economy can be really sustainable only if it is based wholly on renewable resources, which we consume at a rate not higher than their rate of regeneration or replenishment. Logically, then, a sustainable economy cannot be an industrial one as it is today, or even as it was in any earlier epoch, for an industrial economy is based mainly on non-renewable resources. With a little less rigour, we may say that a sustainable economy is one that is based *mainly* on renewable resources and uses non-renewables only when absolutely necessary. Exceptions can be made in the case of coal, iron, and aluminium. Coal will last a few more centuries, and iron and aluminium are truly abundant in the Earth's crust. Moreover, these two metals are highly recyclable.

One cannot simply say that renewables will take over when non-renewables have become inaccessible. We have seen that solar electricity is not viable. And viable renewables will surely not be available in the same enormous quantities as non-renewables plus renewables today. Unless population decreases, per capita availability of resources will be much lower than today. But we see that the global population continues to grow. The most important contribution of the third world countries towards global sustainability should, therefore, be an early end to population growth, while that of the industrial countries should be a drastic reduction in total resource consumption.

Protagonists of a technological efficiency revolution are trying so hard to raise the resource productivity or, in other words, lower the

MIPS of everything that gives a service. But there are many simple ways of doing that. Anybody who has seen how a house is built in India would agree that the MIPS of such a house is very low compared to a German house. The secret of this Indian success is the high labour-intensity of the process. It is actually a double success, for this process also contributes towards mitigating unemployment. There are many other simple ways of reducing MIPS. One can insulate one's body against cold with three pullovers instead of insulating the whole house with high-tech super-materials and super-glass windows. It is better still if the pullovers are hand-knitted, as they are in India. Fifteen families can jointly own a pool of three cars and use them when needed, instead of each family owning a hypercar (high-tech, highly fuel-efficient car). This is practised to some extent in Germany and is also recommended by Weizsäcker *et al.* (Weizsäcker *et al.* 1997: 128–30). We can wear a shirt as long as possible instead of trying to invent a technology to make more shirts with less cotton.

In discussions on sustainability we should not use the terms 'sustainable economy' and 'sustainable society' interchangeably. The former is an essential condition for the latter, but society is more than the economy. A society may collapse or suffer too much even before the economy has crossed the limits of environmental and resource sustainability. This may happen through aggression from outside, civil war, riots, revolts, growth in criminality, chaos, disorder, and so on. All these may result from too much poverty, unemployment, exploitation, and oppression, all of which are already there. Our goal should therefore be a sustainable *society*. This general goal consists of five more concrete goals: the economy must be made sustainable; acute poverty must be overcome/prevented; all able-bodied people must be meaningfully employed; social security must be guaranteed for those who are too old, too young or too ill to work; social and political equality must be guaranteed and economic inequality reduced to a tolerable level. How much and what kinds of freedom and democracy would be possible in the framework of attempts to attain these goals would depend on how easy or difficult they would be to attain. We would have to accept Hegel's dictum: 'Freedom is the recognition of necessity'.

The work to attain these goals must be taken up simultaneously in all societies, globally. For war between neighbours may break out due to troubles in one society. Sustainability or social peace achieved in one society may be ruined because of an influx of problems from other societies. This would obviously mean that exploitation or oppression of one society by another must be overcome. This work is very urgent. We cannot wait for long. Societies are breaking down in front of our eyes – in the former Soviet Union, the former Yugoslavia, the former Somalia, Afghanistan, Rwanda, Burundi, Liberia, Algeria, and so on – causing people immense suffering.

NOTES

1. Collectors are principally photovoltaic (PV) cells or aluminium mirrors, but can be anything that collects sunshine to produce electricity or heat.
2. Of course, a large part of manual labour is nothing but energy, but for the sake of convenience, it is better to reserve the term 'energy' for non-manual sources of energy.
3. Actually, it seems to me that only the printer wears out soon, on average in four years. Other parts of the computer probably only become obsolete in four years. Their real life-span may be somewhat longer.
4. 'Environmental space is the total amount of pollution, non-renewable resources, agricultural land and forests that we can be allowed to use globally without impinging on the access of future generations to the same amount. The amount of environmental space available is, by definition, limited' (Brakel & Zagema 1994: 7).

HAPTER 5

Eco-Capitalism – Can It Work?

Despite all the facts and arguments presented in chapter 4, most economists continue to believe in growth. They say there need not be any conflict between ecology and industrial economy. Some even believe that ecological measures can bring net *economic* gains for industry and trade. In the name of sustainable development, the Brundtland Report gives a clear call for 'more rapid economic growth', including in the already highly industrialised countries (WCED 1987: 89). Surprisingly, many reputed environmentalists believe the same. For example, Friends of the Earth Netherlands write that in their scenario for a sustainable Netherlands, 'economic growth will also continue to be possible …' (Brakel & Zagema 1994: 8). Even Arne Naess, the deep ecologist, is reported to have personally endorsed sustainable development (cf. Orton 1994: 13). They are all wrong. It is none the less worth examining how they propose to attain their goal(s). Who knows, perhaps some of their ideas are useful for attaining the five goals set out at the end of chapter 4?

I have argued earlier that one should desire a socialist society not so much because it is or it could be more efficient than capitalism, but because socialist values are superior to capitalist values. For this reason alone one could reject any capitalist policy. But if the set of policy proposals usually known as eco-capitalism can help to attain at least some of our five goals, and no other set can, then we have to accept eco-capitalism.

I shall now give a short summary of the theory and broad outlines of eco-capitalism, followed by my comments and criticism.

ECOLOGISING MARKET FORCES

For protagonists of eco-capitalism, 'the elegance and virtue of free markets' is that the price mechanism 'signals to consumers what the cost of producing a particular product is, and to producers what consumers' relative valuations are' (Pearce *et al.* 1989: 154). They concede that the market cannot take care of everything. They also speak of market failures. They concede that the free market, especially its underlying principle of self-interest, is largely responsible for the ecological crisis, because self-interest impels enterprises and individuals to externalise costs. But they believe that the economic instruments and mechanisms available within a capitalist market economy, especially the price mechanism, are the best means of solving the ecological problem. Capitalism has other mechanisms for the purpose, but most theorists of eco-capitalism say that the main mechanism used until now, namely regulations to control environmental behaviour, has proved inadequate.

Regulation has failed to produce satisfactory results for the following reasons: firstly, measures to protect the environment are a cost factor. It is therefore rational for an enterprise to try to avoid them. This is also valid for individual consumers, for whom such measures may cost not only money but also sacrifice of comfort, convenience, and flexibility. In the face of this strong, 'rational', negative motivation, regulations have little chance of being sincerely adhered to. Alternatively, the government must spend millions on the machinery of enforcement.

Secondly, even with full adherence to regulations, the result may be less than what is possible. Lutz Wicke gives the following example from the German context: once an enterprise gets the permission to emit, say, 20,000 tons of nitrous oxides, it can continue to do so during its factory's whole life-span. Of course, if the technology has in the meantime developed further, the authorities may try to compel the enterprise to reduce its emissions to, say, 10,000 tons. But the enterprise can argue that the extra costs would be an unbearable financial burden, or that the new technological development cannot be used in its factory for technical reasons, none of which can easily be disproved. In any case, the enterprise would try to delay the execution of the order (Wicke 1988: 158).

Thirdly, combining regulations with moral appeals can be ineffective. In Germany, car owners who forgo using a car may lose two hours' leisure per day, while reducing pollution caused by cars by only one twenty-millionth (ibid: 155). The disproportion between private loss and public gain is demotivating.

Fourthly, there is the free rider problem, that is to say, car owners who continue to use their cars benefit from the self-sacrificing act of those who forgo. This demotivates people who might otherwise respond to moral appeals.

For these reasons, the protagonists of eco-capitalism favour economic instruments as the main ones. Taking measures to protect the environment, they say, must be in one's own economic interest. If the enterprise in Wicke's example sees that it can reduce costs by reducing its emission to below the permitted 20,000 tons, then 'it is no longer interesting to tell the authorities why a particular measure is allegedly or really impossible for technical or financial reasons. Then the management would spur the engineers to search for economic ways to solve the pollution problem' (ibid: 160).

Their approach towards bringing about this change has been expressed in three different formulations of an identical aim: (a) environmental costs that have always been externalised must be internalised; (b) polluters must pay for the pollution; (c) prices must tell the whole truth about costs. Their approach operates mainly through the price mechanism in a free market, but government and legislators have to play an important role.

There are many economic instruments available to attain their aim, of which I shall describe only the main ones. The logic of eco-capitalism can be best understood if we take a very radical version of it. Herman Laistner (1989: 132 ff) demands that all taxes and charges on labour and income from labour (except on income from the labour of others), which ultimately the employers have to bear, should be abolished and replaced by ecological taxes and charges on all products, on all uses of natural resources, and on all demands made upon the environment such as the capacity of nature to absorb pollution (its sink function), the beauty of a landscape, or the health-bestowing quality of a place. At the same time, all subsidies and allowances should be abolished. The logic is clear: since all products, through the processes of both production and waste disposal after consumption, adversely affect the environment, their prices should internalise – that is, fully reflect – the costs of preventing or repairing the damage. Since in the ultimate analysis it is the consumers who pollute – because their demands are responsible for all production – they should pay all the costs.

Examples of ecological taxes and charges might be: a tax on all built-over land; a charge on any quantity of water used (except rainwater) – charges on water used by industry would be determined by the degree and kind of environmental change caused; taxes on the use of minerals – these should be high enough to cover all additional costs consequent on their extraction or recycling; taxes on the burning of fossil fuels – in the case of nuclear power, a tax to cover costs of security and disposal of wastes; charges on all kinds of harmful emissions – in the case of immissions (emissions from across a border), special import duties on goods from the country concerned.

The consequences of such tax reform would be twofold. Firstly, the prices of natural resources, environmental services, and all products would rise. Producers would therefore have an incentive to use less of them in order to reduce costs. Individual consumers would feel compelled to use them frugally. This would lead to conservation of resources and effective environmental protection. Secondly, because labour would become cheaper, employers would be inclined to employ workers rather than machines. This would contribute towards solving the unemployment problem.

Not all protagonists of eco-capitalism are so fundamentalist. Ernst Ulrich von Weizsäcker, for example, takes seriously the criticism that eco-taxes would give rise to a new 'tax jungle'. In the German context, he proposes only 10 new taxes and wants to abolish 200 old ones (Weizsäcker 1989: 172).

Some protagonists of eco-capitalism have drawn attention to a disadvantage of ecological taxes and charges: while some reductions in the (rate of growth of) consumption of natural resources, in demands on environmental services, and in production of goods and services would result, the state would have no control over how great the reductions would be. As an alternative that would allow the state to lay down and control quantities, they propose tradable permits (licences). In the case of pollutants, the state can fix the maximum annual permissible quantity for the whole economy, and pollution permits can be granted, sold or auctioned to those who need them. Enterprises that reduce emissions to lower than their permitted levels can sell their surplus permits at a profit to those who cannot bring emissions down far enough. Even if no profit can be made in this trade, it can give enterprises a degree of flexibility from which the national economy as a whole would benefit. If, over the years, the state were gradually to reduce the permitted quantities for the whole economy to very low levels, then the permits would become very costly and the incentive to reduce emissions would be correspondingly high. Technologies and products that generate high emissions would gradually be pushed out of the market.

What is possible with emissions is also possible with solid wastes and resources, both renewable and non-renewable; in case of the non-renewables, one should speak of depletion permits. It is also possible with regard to population size; in fact, Daly speaks of a 'marketable birth licence plan' (Daly 1977: 72).

Protagonists of eco-capitalism aver that they do not want the state to use eco-taxes to increase its revenue, but that the additional revenue raised through them should be given back to the economy in some form. The same applies to revenues that the state might earn through selling or auctioning tradable permits.

Another economic instrument that would generate a strong incentive for enterprises to take anti-pollution measures is to increase their liability for all environmental damages caused by them. Enterprises would insure themselves against claims arising from such damages, but the premiums they would have to pay would be reduced if they could show that their plant, products, and production processes contained less risk of causing damage than others'.

One thing must be made clear. When the state or enterprises sell or auction tradable pollution permits, or when visitors have to pay to enjoy the beauty of a landscape (a national park), then even those parts of the environmental services that have always been free goods would get a price tag and be treated as commodities. Protagonists of these instruments say without hesitation that it should be so. Pollution charges also transform environmental services into commodities, for an enterprise can then legally pollute more by paying a higher charge.

ECO-KEYNESIANISM

Another current of thought within eco-capitalism can be called eco-Keynesianism. Its protagonists also want to solve both the ecology problem and the unemployment problem. But they do not think that relying mainly on market forces can produce this result. For example, Rudolf Hickel (1986) points out that the 'polluter pays' principle cannot – either in the form of tradable pollution permits or that of eco-taxes – offer a solution to the problem of old, unregulated toxic wastes dumps, which is a big problem in all advanced industrial societies. To solve it, according to eco-Keynesians, large-scale, labour-intensive state action is necessary. They also advocate the further development of eco-technologies and of an eco-industry, which would absorb hundreds of thousands of unemployed people. The state must actively promote this development by creating demand through its own projects.

With regard to current environmental problems, Hickel thinks the policy of an increased role for the market to be inefficient. In the case of tradable pollution permits, firstly, the instrument does not itself lead to a reduction in total pollution; the state must reduce the total figure of permitted pollution from time to time. Secondly, an ecologically conscious entrepreneur or some eco-groups could buy up and hoard pollution permits in order to prevent others from using them, or a large company could buy them up in order to force its competitors to reduce production. Even without these motives, the normal operation of market forces may lead to the concentration of pollution permits in the hands of a few large companies. In order to prevent such things and to allow smaller companies to get the number of permits they require, a detailed system of regulations, fines, and punishments would be necessary, which could be created and operated only by the state. A large apparatus must be created to inspect whether the regulations and the permitted limits are being adhered to. After all, there would be limits for a large number of pollutants. Thirdly, for particular regions affected by a very high concentration of pollutants, special regulations would be necessary. During smog alarms, pollution permits would lose their value. Fourthly, the instrument of pollution permits concentrates only on production processes. It gives no incentive for innovation in eco-friendly consumer goods (end-products), such as cosmetics or furniture free of harmful substances. For this purpose commands and prohibitions are necessary. For all these reasons, argues Hickel, it is better to organise environmental protection directly and politically instead of indirectly through market forces. This would also allow governments to differentiate between old and new factories and to fix different limits for two factories in the same branch of industry.

Eco-Keynesians demand massive state expenditure on environmental problems, and a general ecological restructuring of industrial society, in which the state should play the designer's role. The first idea is clear. For instance, investments in sewage treatment plants would both create employment and make the rivers cleaner. This was formulated as a demand by the German federation of trade unions (DGB) in 1985. The DGB demanded that the state invest DM50 billion up to 1989 in 54 ecologically relevant measures. This was expected to create several hundred thousand jobs (DGB 1985). As to the second idea, new, ecologically efficient technologies should replace older technologies. The need for such replacements is great in every industry. Many end-products, also, should be replaced with eco-friendly, long-lived products. So great demand potentially exists. The ecology industry would be not just one branch among many, but a part of all branches. In some cases, whole factories would undergo what is called product conversion. In the case of traffic, the emphasis should shift from private motorisation to public transport; in the case of goods transportation, from road to rail. The state should promote this process.

The instruments eco-Keynesians prefer are ecological injunctions, such as the laying down of limits, flexible charges, orders and prohibitions. They also demand incentives through subsidies to households (say, for better heat insulation), and small and medium-sized enterprises for eco-friendly technologies. Projects that the state itself would undertake should be financed through new taxes, higher rates of existing taxes, and/or reallocating budgetary sums in favour of ecological projects.

CRITIQUE OF ECO-CAPITALISM

Let us now examine whether eco-capitalism can help to attain our five goals.

Inner contradictions: the time horizon

The most serious problem with eco-capitalism is that it is based entirely on the motivating force of self-interest. Its protagonists prefer the term 'eco-social market economy'. But the attribute 'social' means only that the current generations of the unemployed and poor get certain minimum welfare payments. Environmental protection, however, has a strongly futuristic dimension. We are supposed to leave the world ecologically intact and with enough resources for future generations. But how can it be in my interest to do something that would benefit people who would live 100 or 200 years after my death? 'What has posterity done for me that I should do something for it?' This is not a joke, but a logical rhetorical question arising from the assumption that acting only in one's own interest is an economic virtue, good for oneself and perhaps also good for the present-day economy as a whole. In this assumption, one is not even thinking of the welfare of future generations.

This logical problem is also built into the strong faith of eco-capitalism's protagonists in the efficacy of the market-and-price mechanism. Prices in a free market can signal only current relative valuations of living consumers, not those of consumers not yet born. But for correct price formation it is necessary that all potential purchasers and suppliers have access to the market. In the case of goods and services like good land, mineral resources, and the sink function of nature, future generations would certainly be interested in 'buying' them or 'buying' depletion and pollution permits. But they cannot have access to today's markets.

Generally speaking, the time horizon in a market economy is very limited. 'It cannot ... go beyond the amortization time of capital goods, for profitability calculation ... is exactly limited to this time' (Altvater 1986: 100). Of course, big transnational corporations are interested in long-term forecasts which go beyond the amortisation time of capital goods, but for investment decisions they are 'relevant, only if they conform to the competitive situation prevailing at the time in the market' (ibid: 101). This means that the market-and-price mechanism value the present time higher than the future. A top manager of a large German corporation said openly: 'An industrial enterprise cannot work for the next generation. ... We must produce now for the markets and earn money' (*Der Spiegel*, 9 June 1986 : 100f).

So, for the purpose of leaving the world ecologically intact in the interest of the future generations, the market-and-price mechanism is not useful. The state, or society, or community must take care of this purpose on the basis of ethical considerations. Nothing else can work. Protagonists of eco-capitalism themselves sometimes talk of the need for ethical motivation (see, for example, Wicke *et al.* 1990: 149 ff). But self-interest and taking care of the interests of others can be contradictory. In capitalism, ethics can play only an insignificant role, the upper limit of which is observance of the rules of the game and the laws of the state.

Inefficiency and wastage in capitalism

Let us now turn to a concrete problem. A sustainable economy, including the period of transition to such an economy, must be efficient. Wastage cannot be allowed. Admirers of capitalism nowadays concede that with regard to environmental and resource efficiency there is a deficit, which must be removed. But they assert that, apart from that, if allowed to operate freely, capitalism is the most efficient system. In fact, the strongest argument for eco-capitalism is the alleged efficiency of the market-and-price mechanism. But is this argument valid?

Let us for the moment leave aside questions of rich and poor, distributive justice, exploitation, class conflicts, and so on, which also have something to do with efficiency (strikes, for example, are not, economically speaking, efficient). Let us compare the hitherto observed performance of capitalism only with the most common expectations of efficiency from any modern economic system.

The most evident proof of capitalism's inefficiency is large-scale unemployment, which has accompanied it throughout its existence, except in boom periods, although any capitalist society has always had socially useful work waiting to be done. That no unemployed person suffers from hunger in a developed capitalist society does not disprove its inefficiency. For such people get their livelihood from others' work without giving anything in return, while their own capacity to work remains unused – an irrational as well as inefficient system: it is not the best allocation of labour.

Other resources are also wasted: Products that cannot be sold; things that have to be destroyed so that the market price does not fall; built-in obsolescence; irreparable apparatuses; enormous expenditure in advertising, to sell otherwise unsaleable products; trains running half-empty while millions of cars, each transporting only one person, kill and maim thousands in road accidents; superfluous packaging; billions of dollars worth of investments wasted through bankruptcies; workers from Portugal building a few houses in Berlin while German construction workers remain unemployed; the enormous economic damage caused by widespread criminality and vandalism, and the enormous costs of fighting them – these are all examples of wastage, inefficiency and misallocation of resources in a system that is allegedly efficient and rational.

One may argue that inefficiencies and wastage cannot be avoided fully, especially in a free society, and that what matters is whether they are punished, as in capitalism, or rewarded, as in 'socialism'. One can argue that in capitalism an inefficient enterprise would surely be punished through losses or bankruptcy. That is true. But when we compare systems, we have to compare the degrees of efficiency/inefficiency of whole economies and not of individual firms. A German firm that operates in Berlin with cheap, migrant Portuguese workers is, of course, very efficient. But unemployed German workers are evidence of wastage and inefficiency in the German economy. That is the difference between micro- and macro-economics. For ecologists, the latter is more relevant.

One may argue that the *logic* of a system cannot be refuted so simply by pointing at a few inefficiencies and wastages, and that the almost continuous growth in the macro-prosperity of Western capitalist societies would not have been possible if the system had not been on the whole efficient. So let us now look at the logic of the capitalist economic system.

The logic of the system

Waste through overproduction is due not simply to miscalculations and similar human failures. Entrepreneurs work on the basis of price signals from the previous cycle of production and consumption. But they are producing for the next cycle of consumption. Past mistakes can be corrected and the situation can improve if everything else remains unchanged. But in a capitalist market economy, everything is in flux, everything is uncertain. Miscalculations and consequent waste are therefore unavoidable. Experience with past products is of limited use, because new products may push them out.

As far as solving the unemployment problem is concerned, capitalism clearly has a serious debility, for a reserve army of unemployed workers is greatly advantageous to capitalists. It generates a downward pressure on real wages, makes labour easily available and enables capitalists to hire and fire at will (unless checked by law and trade unions).

Of course, welcoming and utilising a situation are not the same thing as causing it. One important factor behind unemployment is, after all, population growth, which takes place even where/when there is no capitalism. But another important factor is inherent in the logic of capitalism. The motive of profit maximisation and the existence of competition constantly drive entrepreneurs, on pain of bankruptcy, to try to invent and/or introduce 'better' technologies, which results in a secular tendency to replace labour with automated machines and computers. This not only gives particular entrepreneurs an advantage over their competitors, but it is also, in the last analysis, *the* general cause of growing labour productivity and hence of the growing prosperity of a society. For the same reason, 'socialist' societies also replaced labour with machines. But in this effort there was no compulsion, it was not a part of the *logic* of 'socialism'. They only desired it. 'Socialist' enterprises that could not or did not automate still functioned and fulfilled plans.

Given this logic of capitalism, it is no surprise that modern technologies, such as microelectronics, are enabling most enterprises to reduce their unit costs and raise profits, while millions of workers are unemployed. The costs of maintaining the unemployed and the poor are to a large extent social costs, whereas the benefits of replacing labour with machines are reaped fully by individual enterprises. The advantages of doing this are so great that no eco-capitalist measure such as that proposed by Laistner can be good enough to motivate capitalists to do the opposite, and replace machines with labour.

It is said that such unemployment is due to high wages, but it is not simply a question of how high the wages are. Unlike a worker, a machine does not get full wages while it is sick, does not demand paid holidays, a bonus, and an old-age pension, does not have to feed a family, does not refuse to obey orders, does not strike, and does not need to sleep at night. In capitalism, labour is not only costly but also a very troublesome factor of production, because workers are humans. The hope that an ecological tax reform would solve the unemployment problem within capitalism is, therefore, unfounded. Moreover, a high wage level in advanced industrial countries is generally presented as one proof of the superiority of modern capitalism. If wages have to be reduced in order to create employment for all, then this proof is no longer there.

All these points belong to a traditional leftist critique of traditional capitalism, but they have to be stated here, because, firstly, they have not become invalid after 'socialism' has failed, and secondly, because eco-capitalism is capitalism. But what about the most important of the five goals? Is it possible to make the economy sustainable within the framework of capitalism?

With the paradigm shift propounded in chapter 1, one aspect of the critique of capitalism has undergone a fundamental change. The critique is no longer that capitalism ultimately fetters the forces of production. On the contrary, the critique today is that it has developed and is developing them so much that they cause massive degradation of the natural basis of many forms of life including human life. Because of this change, the traditional leftist critique of capitalism has indeed become obsolete, but not fully. Capitalism continues to degrade humans physically and psychologically, tends to reduce them to mere money-making machines. Its very logic fetters the higher potential of people and society – potential that does not generate profit. Its basic principles – self-interest, greed, and competition – promote criminality.[1] For this reason, and to this extent, the leftist critique of capitalism remains valid and relevant. It is also highly relevant to the task of transforming present-day economies into sustainable ones. This transformation cannot take place, as even many protagonists of eco-capitalism concede, without an ethical approach, without the readiness to sacrifice self-interest. But degraded humans are not capable of this ethical approach and this sacrifice, as is evident from the obstructive reactions of the majority of citizens of industrial societies to the challenge of this transformation.

There are, however, further reasons why neither a sustainable economy nor a transition to it can function within the framework of capitalism.

Growth dynamics in capitalism

There is a fundamental contradiction between the logic of capitalism and that of a sustainable economy. From the material presented in chapter 4, it is evident that, at least in the advanced industrial economies, a process of contraction is necessary if they are to become sustainable. But in a capitalist economy there is a built-in growth dynamic. There are three causes of this. Firstly, entrepreneurs are not satisfied with simply earning enough for their livelihood. They want to earn much more. That is why they are prepared to take risks, invest their money, and work hard. Secondly, they do not or cannot consume their whole profit. Nevertheless, they want to make more profit in the following year (greed). That is why they invest the greater part of their profit in expanding the enterprise. Thirdly, there is an external compulsion to grow. Capitalists cannot say 'enough'. If a capitalist does not take advantage of economies of large scale, his/her competitors would do so and push him/her out of business. In the capitalist world of brutal competition there is a rule: expand or perish. All try to expand, and the net result is that the economy as a whole expands.

And that is also considered to be normal and good. One does not need a negative growth rate to speak of crisis: a growth rate below 2 per cent is a crisis that results in thousands of bankruptcies.

Almost all protagonists of eco-capitalism believe, in spite of the above contradiction, that ecological modernisation of the industrial economies would promote growth, and that that growth would be sustainable. But if growth in already industrialised economies is unsustainable for very concrete scientific reasons, it cannot become sustainable merely through the adoption of some eco-capitalist policies.

On closer examination, one sees that, for some, this belief is based on a logically impermissible reduction of the concept 'sustainability' to current environmental quality only. Thus Pearce *et al.* write that environmental quality frequently improves economic growth by improving the health of the workforce, creating jobs in recreation, tourism, and so on (Pearce *et al.* 1989: 21). That is true in the short term. But the long-term problem of limits to resources remains. Good health in the workforce is a value in itself. But if it is to improve economic growth, then it must work with resources, most of which, at least in industrial economies, are non-renewable. The same applies to recreation and tourism. Pearce *et al.* also give the examples of propellants (in spray cans) that do not harm the ozone layer, and unleaded petrol (ibid: 22). Here also, total resource consumption will go up, with an adverse impact on the environment, if the improvements are 'integrated into capital investments' for growth.

Of course, by 'sustainability' and 'ecological modernisation', one also means dematerialisation through an 'efficiency revolution'. But, as Fred Luks shows, if resource consumption in industrial societies has to go down in the next 50 years by a factor of ten and, at the same time, economic growth is to continue at the rate of 2 per cent per annum, then resource productivity must rise by a factor of 27 (Luks 1997). Is that a realistic hope?

Growth in benefits or economic growth?

Another cause of the above mistaken popular belief is a confusion between growth in total benefits and economic growth. Wicke *et al.* calculated that towards the end of the 1980s, in West Germany, monetarily calculable economic damages caused by environmental pollution amounted to DM120 billion (Wicke *et al.* 1990: 68). Since, according to them, the costs of investments necessary to avoid such damage would be much less than this, it is clear that measures to prevent environmental damage generate greater benefit than they cost (ibid: 70).

This is, in fact, common sense: prevention is better than cure. The problem, however, is that in capitalism entrepreneurs are not interested in public benefits but only in their own profits, just as they are not interested in the social costs but only in costs arising to them. Their profits come from their sales, and sales increase when the economy grows in the usual sense of growth in GNP. But clean air, pure rivers and seas, the undamaged beauty of a monument – the expected results of investments that they might be directly or indirectly compelled to make – are not products that they can sell. This is why resistance to such investment is sure to arise in capitalism. Entrepreneurs do not behave stupidly when they spend money on advertising in order to motivate people to buy more and more useful or useless or even harmful things.

Wicke *et al.* estimate that damage to health caused in West Germany in the mid-1980s through air pollution cost DM2.3–5.8 billion (ibid: 67). If air pollution is drastically reduced, all citizens will enjoy better health. The investment in preventing air pollution would, of course, generate growth in the industries that produce the required equipment. But since cure generally costs more than prevention, the better health of citizens would cause the medical and pharmaceutical industries, services, and professions to suffer a comparatively greater slump. The net impact on the economic cycle would then be negative.

One may argue that the extra money that citizens would have in their pockets because of reduced costs in health care would be spent on other things. But, firstly, because in advanced industrial economies surplus production capacities prevail and most people are saturated with consumption, the extra money may also be saved. This may lead to the so-called paradox of thrift causing a further fall in GNP.[2] Secondly, if this extra money is spent on other things, then there is no net gain for the environment. The air will have become cleaner at the price of other kinds of negative environmental impact due to increased production of other things. Herman Daly illustrates the point as follows:

> Suppose the government taxes automobiles heavily and that people take to riding bicycles instead of cars. They will save money as well as resources. But what will the money saved now be spent on? If it is spent on airline tickets, resource consumption would increase above what it was when the

> money was spent on cars. If the money is spent on theatre tickets, then perhaps resource consumption would decline. However, this is not certain, because the theatre performance may entail the air transport of actors, stage sets, and so on, and thus indirectly be as resource consumptive as automobile expenditures. (Daly 1977: 62)

Of course, it is possible to increase benefits without causing higher resource consumption. If a friend helps me to cook, I gain some benefit, as does the friend, because he/she enjoys my company. But I do not have more food on the table. If I repair my torn shirt, the benefit of having a shirt is reproduced. In the process, no resources except a little thread have been consumed. The benefit increases if I and my friend use each other's shirts of different colours and designs.

But in capitalism, entrepreneurs are not interested in such growth in benefits, which do not increase their sales. It becomes a little more interesting for them if I ask a tailor to repair my shirt and pay him for the work. A tailor is, after all, a small entrepreneur. But from the standpoint of entrepreneurs in general, it is best if I throw away the shirt and buy a new one.

It is therefore not surprising that in capitalism, GNP is not a measure of benefits, but only of the value of goods and services produced *and* sold. GNP grows when people consume more resources and more paid labour, and throw away goods that can still be used. Only that is economic growth, and only that interests entrepreneurs. In this process both benefit and harm can result.

The world market

Of course, most protagonists of eco-capitalism also write about exports and imports, but they seem to think that their ideas could be implemented without any problem if only their nation state had the political will to do so. They seem to ignore the fact that today's market is a world market. The power of the nation state extends only to its borders, but capital has become very mobile and knows no patriotism. Similarly, many environmental entities – rivers, air, seas – do not respect national territories.

The market tends to globalise economic processes. This has to be so. The development of new technologies that are more efficient and more productive only if used on a larger scale, the growing division of labour, and allowing comparative advantages to play an important role – three essentials for the growth of both profits and prosperity – necessarily lead to growth of the world market. If eco-capitalism is to be realised by nation states, then this globalising dynamic has to be curbed, against the logic of the market. Free movement of capital across borders also has to be curbed. But this will curtail economic growth and prosperity. You cannot have both restrictions on the growth of the world market and further economic growth This contradiction between the globalising dynamic of the market and the limited jurisdiction of nation states was the main reason why Keynesian interventions in the 1970s failed to achieve the expected results. And for this reason, today, eco-Keynesianism will also fail.

Many Greens and environmentalists advocate policies expressed in the slogans 'small is beautiful', 'regionalisation of the economy', 'home market orientation', and so on. They are all good for the environment, but one cannot then let the economy be directed by market forces, which logically tend towards globalisation.

Of course, growth of the world market can also be useful in a certain way for securing a sustainable development path. Pearce *et al.* show it by considering an extreme theoretical example

> in which an economy imports all its raw materials, uses indigenous technology and human skills to convert it to a final product, and then exports the final product. Because it adds significant 'value added' in the process, it can then further import all its food needs as well. The nation's stock of natural resources remains intact, but the nations from which it imports may well be experiencing a decline in their natural capital stock because it is being exported. (Pearce *et al.* 1989: 45)

This is called 'importing sustainability' 'at the cost of non-sustainability in another country' (ibid).

From the real world market they give the example of trade in tropical hardwoods. Japan and the European Community together import 90 per cent of the world supply, 80 per cent of which comes from only five countries: Malaysia, Indonesia, the Philippines, the Ivory Coast and Gabon (ibid). It is well known that tropical forests are being logged unsustainably.

Of course, Pearce *et al.* think that it is possible for the countries that 'import sustainability' to compensate the countries that are 'exporting sustainability' by giving 'aid for sustainable development' (ibid: 47). But we know that because or in spite of foreign 'aid', the external indebtedness of most third world countries is continuously growing and becoming more and more unbearable. And we know that partly for this reason, their economies are becoming more and more unsustainable.

It is too easy to say, as most leftists do, that unequal exchange is the cause of growing unsustainability in the third world. It can easily be shown that even benevolent foreign aid and equal exchange between trading partners would not make growing world trade sustainable for all countries. In fact, the problem is very basic and very simple to understand. In a world with finite resources and nature's finite capacity to absorb pollutants, only a few countries can enjoy the average standard of living prevailing in the USA, Germany or Japan. All countries cannot develop, let alone develop in a sustainable manner, to such levels. Genuine foreign aid and equal exchange could perhaps help promote sustainability, especially in the third world, but only if, simultaneously, the total volume of world production goes down. But it is continuously growing, driven by world market forces. Equal exchange in today's world market would redistribute prosperity. It would also redistribute unsustainability, but not reduce it globally.

Ecological tax reform

Let us now examine the economic instrument most popular among protagonists of eco-capitalism: ecological tax reform.

The more labour-intensive a technology, the smaller the quantities of resources it requires. Therefore, other things being equal, such technologies cause less pollution. So why were they given up? The answer is simple. To take an example, in the 1960s, West Germany, in spite of full employment, was producing less in the way of goods and services than today, when more than 4 million are unemployed in the same part of Germany. Increasing labour productivity meant increasing production, higher wages, a higher standard of living. In contrast, higher resource productivity would mean more person-hours per unit of production, hence increased labour costs, lower wages and a lower standard of living. To illustrate the point: to renovate an old house or to repair a watch is to avoid waste of resources; but they are labour-intensive. You cannot, generally, have both higher resource productivity and higher labour productivity.

But protagonists of eco-capitalism want both. Laistner, for example, rightly differentiates between GNP and social benefit. He says that the former should fall while the latter should rise (Laistner 1989: 131). However, he also says that if his ideas were implemented, GNP would fall, but net wages, purchasing power, and standard of living would not change (ibid: 134). Laistner is speaking of all people. But how can people's purchasing power remain unchanged if GNP falls?

All advocates of eco-taxes say that the extra revenue earned through them should be given back to the economy – *inter alia*, in order to make labour more affordable. Is it possible? Let us go through a rather simplified arithmetical exercise. Let us take Germany and suppose that petroleum is its only source of energy. Let us further suppose that at present German industry and private citizens are consuming 1,000 litres of petroleum per year and that the price is DM1 per litre, of which DM0.30 is tax. We get the following result from the present situation (case 1):

amount consumed	price per litre	of which tax	revenue for state
1,000 litres	DM1	DM0.30	DM300

The DM300 revenue is already being used for various purposes.

Let us now suppose that it is necessary to reduce total petroleum consumption to 500 litres. Let us also suppose that for this purpose the government raises the tax by another DM0.30 per litre and attains its goal. We then get the following result (case 2):

amount consumed	price per litre	of which tax	revenue for state
500 litres	DM1.30	DM0.60	DM300

But now suppose that the Germans reduce their petroleum consumption to 400 litres. We then get the following result (case 3):

amount consumed	price per litre	of which tax	revenue for state
400 litres	DM1.30	DM0.60	DM240

In case 2, the goal has been attained, but there is no extra revenue that can be given back to the economy. In case 3, there is a revenue loss.

My simplified arithmetical exercise is corroborated by at least one concrete case. In 1993, the public utilities of Frankfurt made a record loss of DM195.7 million. They had expected a loss, but of only DM90.1 million. The Green–Social Democratic coalition administration said this heavy loss was due mainly to reduced water consumption, an effect they had themselves sought through their double step of raising water prices and a public campaign against wastage (*Frankfurter Rundschau*, 24 January 1994). You cannot eat the cake and have it too.

If petroleum consumption fell only to 600 litres, we would then get the following result (case 4):

amount consumed	price per litre	of which tax	revenue for state
600 litres	DM1.30	DM0.60	DM360

The state's revenue has increased by DM60. But it has failed to reach the target.

In this arithmetical exercise, an increase in the state's revenue from the tax is only one possibility among three. But Michael Jacobs maintains that the state would certainly gain some revenue from such taxes, which it could give back to the economy. Let me quote his argument:

> It might be thought that there wouldn't be any [revenue], since the purpose of the tax is to reduce the pollution (or resource depletion) to the acceptable level, and at that level no tax would be payable. But ... even when the acceptable level is reached there is still a revenue, because the tax is paid on all the damaging activity, not just the proportion which exceeds that level. (Jacobs 1991: 146)

The difference between cases 2 and 3 above and Jacobs' assertion, which is correct, can be easily explained. We are asking different questions. Jacobs is asking whether there would be *any* revenue. I am asking whether there would be any *additional* revenue if the tax were increased. If it can be assumed that the tax Jacobs has in mind did not exist previously, then, of course, this particular tax would generate some revenue, for it is very unlikely that some particular pollution or resource depletion would fall to zero as a result of the new tax.

Jacobs' assumption is rather unsound, however, because in modern industrial countries there is hardly anything left on which consumers do not already have to pay, directly or indirectly, a tax or a charge. But let us, for the sake of argument, accept what Jacobs says. It is somewhat similar to case 4 in my arithmetical exercise. Let us examine it more closely.

As we know, the state raises many other taxes from the economy: value-added tax, sales tax, corporate income tax, business tax, and so on. In the petroleum industry, production has fallen to 60 per cent of the previous level. So revenue from these taxes in this industry has also gone down proportionately. The same is likely to happen in the branches of industry which supply machinery and other things to the petroleum industry. The combined revenue loss to the state in these two areas may be much higher than DM60.

Here we see the difference between microeconomics and macroeconomics. Although revenue from a particular tax may increase, total revenue from all taxes may decrease. Total revenue from taxes is always a share of GNP. If GNP falls as a result of a drastic reduction in resource consumption (think also of the multiplier effect), then it is very unlikely that there would be any increase in the state's total revenue, which it could give back to the economy.

Particularly, a new tax (or an increase in an existing tax) on energy would have a very negative effect. Since the industrial revolution, the use of energy in place of human and animal labour power has been the most important factor in increasing labour productivity. If the use of energy is drastically reduced, then, of course, unemployment may go down, if more people are employed to replace energy and machines. But because of greatly reduced labour productivity (while resource productivity goes up), prosperity would surely go down.

But argument and logic have no effect on the quasi-religious faith of most protagonists of eco-capitalism in the miraculous efficacy of the instrument of ecological tax reform. Thus, unlike Laistner (who appears to be only a little less faithful), Loske *et al.* believe that, through their ecological tax reform, a country can become also 'economically, i.e. monetarily, richer' (Loske *et al.* 1996: 186). Jacques Delors, the former Chief of the European Commission, his successor Jacques Santer, and even a British Conservative politician believe that they can promote economic growth and employment through ecological tax reform, and Weizsäcker *et al.* (1995: 226) quote them as witnesses.

Let us suppose that, in spite of revenue losses in other areas, the state gives the DM60 (in case 4) back to the economy. A part of it would go back to private citizens. But if, say, car-owners get back the money they have paid in the form of an increased tax on petroleum via, say, a reduction in income tax, then they need not reduce car travel at all. If a part of this money is given to non-car-owners as a bonus (as suggested by many), then these people can more often use a taxi with their extra money. The environment would doubtless become better if the state could be persuaded to use all of this money to promote public transport, but since the latter must also pay the higher energy price and cannot reduce its services, its net monetary gain (and that of its users) may be, at best, zero.

The state could probably be persuaded to give part of the DM60 back to industry, in the form recommended by most protagonists of eco-taxes, namely reduced social insurance contributions. Would this be enough to induce industry to employ more workers instead of machines? I guess not. In industrial societies, wage rates are very high, which is one reason why machines replaced labour in the first place. A little reduction in incidental wage costs would not suffice to reverse the trend.

If real wages could be reduced sufficiently, then, of course, industry's production costs would not go up owing to higher numbers of workers. But that has nothing to do with current proposals for an ecological tax reform.

The all-important question is what the government wants to achieve. So far, all (increases in) eco-taxes that have been introduced anywhere have been so mild that they have indeed increased the state's revenue without causing any fall in GNP. In some cases, they may have slightly reduced the intensity of particular forms of pollution. But no real overall improvement is possible without an overall reduction in materials flow. To take a concrete example, in 1979 the Swedish government introduced a carbon dioxide tax on petrol. But the initial effect of reducing carbon dioxide emissions was nullified by growth in traffic volume (Dudde 1996: 27). Summing up a similar experience with such a tax, the Danish department of environment says that 'for an effective blow against motor traffic, a price would be necessary that would amount to at least 20 Krone (DM5) per litre instead of the 6.50 Krone (DM1.70) today' (*Frankfurter Rundschau*, 16 October 1998).

Let us now look at a study undertaken in Germany at the instance of Greenpeace. The Deutsche Institut für Wirtschaftsforschung (DIW) in 1994 estimated the probable consequences for the German economy of an ecological energy tax imposed in Germany alone. The tax rate was to rise gradually up to the year 2010. The DIW concluded that such an energy tax would have no negative effect on the national economy, and that the necessary structural change in the economy could be made 'without prejudicing economic growth and welfare' (DIW 1994: 18) The DIW forecast that, in comparison to 1990, energy consumption would fall by 7 per cent by 2005 and by 8 per cent by 2010 simply as a result of technological progress; under the influence of the ecological energy tax, it would go down by a further 11.6 per cent by 2005 and a further 14 per cent by 2010 (ibid: 14).

That increases in energy costs and consequent reductions in energy consumption would not have any adverse effect on economic growth and welfare is difficult to believe. On closer examination, one finds that this hope is based on the naive assumption that energy-intensive industries would remain in Germany despite rises in the price of energy in this country alone. Decisions of industrialists on this question would, of course, depend on whether the energy price rises would be felt as mild or steep, bearable or unbearable, a question that cannot be answered today. It is, however, worth noting that another study, by the Rheinisch–Westfälisches Institut für Wirtschaftsforschung (RWI) in 1996, makes the more realistic assumption that production of energy-intensive goods – such as steel and basic chemicals – would be transferred to other countries and would cause a loss of 413,000 jobs, which would not be compensated for by the creation of new jobs in other sectors (RWI 1996: 25).

The DIW forecast that revenue from the energy tax would be DM9 billion in the first year (1995), DM120 billion in 2005, and DM205 billion in 2010 (DIW 1994: 14). These sums would be returned to the economy. Of course, if the energy-intensive industries were to remain in Germany and economic growth continue, then extra revenue might result. But many protagonists of ecological tax reform have realised that the RWI's assumption is more realistic. Weizsäcker now says that the energy that industry needs for production – the process energy – must be exempt from the proposed ecological energy tax (Weizsäcker 1996). (The government of Denmark had to make such concessions to industry.) If this advice were followed, then the extra revenue from this tax would be very much lower. Then it would be wrong to assume that much could be given back to enterprises to lower their total wage costs. But this very assumption is the basis of the DIW's hope that lowered total wage costs would enable relatively labour-intensive sectors to reduce the prices of their products, which would then lead to greater demand and increased employment in these branches. The DIW estimated the effect to be 600,000 additional jobs in 10 years.

The DIW has considered only an ecological energy tax, but Weizsäcker proposes ecological taxes on all raw materials (ibid). This would create more problems for German industry if its competitors do not suffer such taxes.

Let me finish this section with another example of naivety on the part of protagonists of an ecological energy tax. Some researchers calculated that in Europe, the various constituent parts of a pot of yoghurt travel 8,000 km before the yoghurt reaches the consumer (Böge & Holzapfel 1994). The argument is that if the price of fuel were raised sufficiently, then this 8,000-km travel, which is absolutely unnecessary, would be reduced to a very small figure, and consumers would still have their yoghurt.

This is quite right. The travel is not at all necessary. But (industrialists are not stupid) it is profitable, especially in capitalism. Since the days of Adam Smith we have known that the growing division of labour is one of the secrets of growing productivity and the growing wealth of nations. Eight thousand kilometres of transportation is necessary to optimise the productivity of the various factors of production, and to get the benefits of economies of scale and the comparative advantages of different places of production. Of course, if the absurdities arising from the rules, regulations and subsidies of the European common market could be removed, the transportation of goods in Europe could be reduced to some extent. But the principle remains: growing division of labour necessitates growing transport. It is not for fun that Germany exports about one-third of its production and imports a corresponding amount of goods. Localising production is in the interest of the environment, but it is in the interest of neither capitalists nor consumers, because prices would go up. These laws of the capitalist and/or industrial mode of production and consumption cannot be violated even by so-called progressive capitalists if they do not want to go bankrupt in a competitive world.

Where will the money come from? – Dilemmas of eco-Keynesianism

Suppose the state does earn some extra revenue through eco-taxes. Should it be given back to the economy? In ecological literature we often read of 'our debt to nature'. If we pay some money back to a creditor, we cannot also give it to another person. We can pay back our debt to nature in two ways: generally produce and consume less and let nature recuperate, as far as possible, and/or repair the damage inflicted, if it is reparable. In either case, there would be no money to give back to the economy, not in the sense meant by the protagonists of eco-taxes. No state has yet taken the first route. But the second has been accepted in principle.

'Our debt to nature' is a metaphor for accumulated past pollution and other damage inflicted on nature, which should be removed or repaired and will cost a lot of money. It can also include the enormous expenditure necessary to prevent current and future pollution and damage. Eco-Keynesians demand that the state itself fulfil part of these tasks with tax money and, for the rest, motivate or compel businesses to invest large sums replacing old technologies with new, eco-friendly ones. The state should, according to them, also subsidise eco-friendly measures. But where will the money come from?

In the growth paradigm, this question is answered easily: from economic growth. This has been well formulated by a leading German Green politician, Joschka Fischer, who writes: 'A politics of ecological reconstruction is dependent on mobilisation of enormous sums of money, requires, therefore, a flourishing economy and a financially strong state, so that both can invest in ecological reconstruction' (Fischer 1989: 110). Of course, in capitalism there are also recessions. But Keynesians believe that deficit financing can tackle the problem.

I think that all this is totally unrealistic. Haven't we already seen the failure of Keynesianism? Even states that are not suffering recession, but only experiencing a low rate of growth, are undergoing a general crisis of state finances, which they are trying to tackle at the cost of the poor, that is, by scaling down or dismantling the welfare state. How can they undertake massive expenditure to protect the environment?

Fischer and his like are not even aware of the dilemma contained in the question. The authors of the German Green Party's programme, 'Restructuring Industrial Society' want, on the one hand, to stop degradation of the environment and exploitation of the third world. On the other hand, for their proposals – in essence, the reallocation of money from one budgetary item to another – they take the size of German GNP and state revenues at the time as the basis, although they know that a large part of these are generated through degradation of the environment and exploitation of the third world (Die Grünen 1986).

One particular hope of many eco-Keynesians is worth noting and criticising: that their own national economies would flourish through the export of eco-technologies. For example, the German federation of trade unions (DGB) wrote in 1985 in its major position paper entitled *Protection of Ecology and Qualitative Growth*:

> Development of modern and economical technologies for the protection of environment can open up and secure new markets and competitive advantages. These markets should be seen as the markets of the future, for also in the world market there will be ever stronger demands for products and production processes agreeable to ecology, *since all over the world environmental pollution will increase*. (DGB 1985: 13; emphasis added)

The authors are so blind in their eagerness to make money and create jobs by conquering a new market that they do not even realise that success in this kind of ecological export policy requires the continuation, even exacerbation, of the very disease they purport to cure.

This kind of blindness is apparently prevalent among all kinds of protagonists of eco-capitalism. Laistner expects that Germany would – thanks to the superior grey cells in the brains of her scientists, engineers, and technicians – conquer the market for 'very intelligent products' and 'high-tech machines and end-products', which are all labour-intensive. The market for 'inferior' 'junk goods' he wants to leave to others (Laistner 1989: 138). He does not realise that success in this policy requires that those others remain underdeveloped. But the others will not remain underdeveloped. Whether this kind of nationalistic policy would function in the era of global economy is a moot point. But they certainly go against the declared spirit of sustainable development.

STEADY-STATE CAPITALISM?

It is very strange that many ecologists believe that economic growth could continue in spite of a drastic reduction in resource consumption. It is stranger still that they believe this to be possible within the framework of capitalism. I shall now examine some positions and ideas of such ecologists.

When resource consumption has been reduced by a factor of ten and the optimum combination of labour- and resource-productivity has been reached, we shall have come very close to a steady-state economy

(SSE) at a much lower level than today. In the transition period, the economy would not grow but contract.

For the transition to an SSE and for its future functioning, Herman Daly suggests three institutions which build on the existing bases of the price system and private property. Two of them – transferable birth licences to stabilise population, and depletion quotas to stabilise the stock of physical artefacts and keep throughput of resources within ecological limits – are tradable. The third institution – which Daly calls the distributist institution, consisting of maximum and minimum limits to personal income and a maximum limit to personal wealth – has little to do with the market. The state would issue the birth licences and depletion quotas and fix the limits to income and wealth. The rest would be left to the market (Daly 1977: chapter 3).

I have many doubts about the feasibility of Daly's scheme. Of course, I share his belief in the inevitable necessity of a steady-state economy based mainly on renewable resources. But can it be achieved and can it function within capitalism?

Daly believes in the efficiency and 'discipline of the market'. Markets, he believes, would ensure that the quotas are allocated efficiently. Moreover, although he is advocating social control, he wants to minimise the sacrifice of personal freedom. He aims for macrostability through the quotas and microvariability through the market. The micro is for him the domain of indeterminacy, novelty, and freedom. It should retain the capacity for spontaneous co-ordination, adjustment, and mutation. He writes: 'We should strive for macrocontrol and avoid micromeddling. ... We lack the knowledge and ability to assume detailed control of the spaceship, so therefore we must leave it on "automatic pilot", as it has been for eons' (ibid: 51).

These are standard arguments for a market economy – beautiful literature, but not convincing. Firstly, one cannot speak of discipline if one also admires spontaneity, indeterminacy, mutation, novelty and freedom. The market 'disciplines' only *individual* entrepreneurs and only in the sense that they must remain competitive. It does not compel them to pay attention to social objectives and the welfare of fellow citizens. Historically, the system is notorious for its anarchy, crises, bankruptcies, inflations, recessions, depressions, and unemployment. Of course, capitalism has also generated economic growth, up to the present level of prosperity of industrial societies. But Daly has turned his back on the growth paradigm. Can the system go on functioning in the context of a contracting economy heading towards a low-level steady state?

Daly's metaphor of an 'automatic pilot' is misplaced. The right metaphor in the case of a capitalist market economy is Adam Smith's mysterious 'invisible hand'. An automatic pilot is a conscious creation based on scientific knowledge and design, and it steers the spaceship predictably in conformity with the trajectory fixed by its creators. It is closer to an instrument in the hands of planners, like a computer. Moreover, unlike markets in the broadest sense, the capitalist market economy, which Daly is here talking about, has not been there for eons, but is of recent origin.

In his book written with Cobb about 20 years later, Daly uses the metaphor of a language and its grammar (Daly & Cobb 1990: 44). This metaphor is also misplaced. It is true that a language was not designed by any one person, but its vocabulary and grammar are written down in books; anybody can master a language. But nobody can master the non-existent 'grammar' of a free market economy. It is neither written down nor visible. That is why, despite many management schools, even experienced businessmen make losses, 100-year-old firms go bankrupt, and whole economies experience periodic recessions and crises.

An important 'design principle' of Daly's SSE is to maintain 'considerable slack' ('margin of error') between the actual environmental load and the maximum carrying capacity. That is right. For the satisfactory functioning of any kind of economy, sustainable or unsustainable, some slack is necessary. One of the causes of malfunction in the former Soviet economy was lack of sufficient slack between production targets and resource flow. But it is easier to have considerable slack in an affluent society than in a non-affluent one. How would it be possible to maintain the necessary slack in a contracting or low-level steady-state economy based mainly on renewable resources? In such an economy, the budget would be so tight that society would not be able to allow itself 'considerable' slack. Daly sees the consequence: 'The closer the actual approaches the maximum, the less is the margin for error, and *the more rigorous, finely tuned, and micro-oriented our controls will have to be*' (Daly 1977: 51; emphasis added). Well, then there would be no chance for a free market economy! This is advocacy of comprehensive planning.

Of course, 'considerable slack' is theoretically possible even in a low-level SSE, if the population drops to the required low figure. But Daly is also a realist. That is why, in regard to depletion quotas, he suggests only a gradual reduction, beginning with the present rate of depletion. How realistic is it then to expect that the population of a society would fall more rapidly than the rate of contraction of its economy? All demographers know how difficult and long-drawn-out a process it would be just to bring population growth to a halt.

Daly and others assume that entrepreneurs would continue to make profits in both a contracting and a steady-state economy. In his SSE, all corporate profits would be distributed as dividends to stockholders (ibid: 56). Weizsäcker *et al.* are even convinced that the profits of entrepreneurs and corporations would increase, the standard of living would rise, and whole countries would become richer if they did what the authors advised (Weizsäcker *et al.* 1997: 200, 202, 204 and *passim*). Daly assumes that the opportunities forgone by the wealthy because of the upper limit to income and wealth, would be available to the less wealthy, who would be paying tax on their 'increased incomes'. He thinks that the guaranteed minimum income could be financed from, *inter alia*, these taxes (Daly 1977: 56). I am not convinced.

What an entrepreneur actually does with factories, machines and workers is to transform raw materials and energy into goods and services, which are then sold at a profit. The ecologists we are considering here demand that in industrial societies materials and energy flow be reduced – gradually, of course, but greatly. Schmidt-Bleek (1994) and Loske *et al.* (1996) demand a reduction by, roughly, a factor of ten. If this reduction takes place, then a proportionately large part of present-day factories, machines, and so on would become superfluous; that is to say, capital in the financial sense would be destroyed. How, then, would entrepreneurs, whose capital has been destroyed, still be making profits? How would a capitalist market economy still continue to function? Would the result not rather be a protracted and accelerating economic crisis until the low-level steady state is reached? Wouldn't that lead, in a capitalist market economy, to chaos and breakdown of the economy and society?

All protagonists of steady-state capitalism argue that since the reductions would take place gradually, entrepreneurs would have enough time to adapt themselves to it. Of course, in the history of capitalism successful adaptation to changed situations has been possible, and it is possible in the future. But it has been and can be possible only in the context of overall economic growth. Destruction of capital has taken place during every economic crisis. This has not led to a collapse of capitalism because after every period of crisis there has been a longer period of economic growth. This way of overcoming future crises and averting a collapse in society would not be available if a policy of contraction were adopted.

Capital never dies a 'normal' death. A machine, say a lorry, has a certain life-span. It works for 10–15 years. After that, the machine 'dies', is discarded. But the capital it embodies never dies. The process of amortisation sees to it that when the old lorry dies, the capital embodied in it continues to live, so that a new lorry can be bought to replace the old one. If a new lorry is not needed because of the policy of contraction, the immortal capital formerly embodied in the now dead lorry becomes superfluous. And it can be safely assumed that much of the plant and machinery working at the beginning of the process of contraction would become superfluous even before the capital embodied in them has been amortised. The conclusion is that entrepreneurs would not be able to adapt themselves to this change. Destruction of a large part of capital in the financial sense would be inevitable. And a large part of the capital that can be amortised cannot be reinvested. A part of it probably cannot even be consumed away, for the amount of goods and services available for consumption would also diminish until the steady state at a low level is reached. (This was approximately the case in the former GDR. People there had too much in savings and too few goods and services to buy.)

All advocates of capitalism know that much of the efficiency and superiority they ascribe to the market depends on the existence of a buyers' market, that is, on the existence of competition between suppliers and the possibility of increasing supply if demand rises (otherwise there would be inflation). But in the context of a contracting economy, the surviving entrepreneurs would enjoy a sellers' market. The gradually diminishing supply of goods and services would face a larger than proportional effective demand. Of course, the income generated would be proportionate to the volume of production. But accumulated past savings and amortised capital, parts of which could not be (re)invested, would inflate the consumption fund of the citizens. The sellers' market would guarantee that even inefficient firms would make profits and even bad quality goods and services find buyers, as was the case in the 'socialist' economies. In a capitalist market economy, this would also result in inflation, which would necessitate state intervention, probably price controls and/or rationing, if the state wants to ensure distributive justice, which Daly and others like him demand.

Only in the primary resources market, as Daly also points out, would there be a buyers' market. Since unextracted stocks would not disappear, and trees would continue to grow, owners of mines and forests would compete to woo industrialists and businessmen, the possessors of gradually diminishing depletion quotas, to buy their minerals and timber.

A central thesis of Weizsäcker *et al.* is that in a capitalist market economy that reduces resource consumption, firms would be able, thanks to an efficiency revolution, to 'earn more by selling less' (Weizsäcker *et al.* 1995: 186). They give the example of some large electricity companies in the USA, which allegedly achieved this marvel. On closer examination, however, it turns out that these companies earned more by selling *other* goods and services, not by selling less of their former sole product – electricity. These companies had partly transformed themselves. They were also selling (or leasing out) equipment (such as energy-saving lamps), consultation, planning, and execution necessary to save energy (ibid: 188). This partial transformation was promoted by the state, in that the official price commission resolved not to allow the electricity companies to make any profit by selling electricity above the figure estimated by the commission to be the real need. At the same time, the commission granted them an extra profit on energy-saving business: they were allowed to claim for themselves 15 per cent of the financial savings a customer made through energy-saving measures (Weizsäcker *et al.* 1997: 159–60).

Another idea from Weizsäcker *et al.* is the tertiarisation of industry. If, for example, manufacturers of photocopiers were to lease out the machines instead of selling them, they would have a strong interest in producing and repairing long-lived machines. They would then continue to make good profit, and resource consumption would go down. A similar result would be obtained if car manufacturers were to rent out cars instead of selling them. The authors report a concrete case from the chemicals industry: Dow Chemicals of Germany, in collaboration with a recycling firm, started renting out a chemical instead of selling it. This chemical is used in industry as a solvent for removing fat, and every given quantity of it can be used up to 100 times. Dow Chemicals and its partner made profits as usual, while resource consumption went down drastically.

These and similar *particular* examples are convincing. But from particular concrete examples of success, the authors draw the general conclusion that the whole business community would prosper if their resource-saving advice were implemented. Car manufacturers who rent out cars instead of selling them, may, of course, make profits in spite of reduced capacity utilisation of their plant and machines. But the fall in car production would surely result in losses in the steel, tyre, and ancillary industries. Similar things would happen in connection with photocopiers and chemicals. In the case of photocopiers, the manufacturers' leasing business would eat up the profits of hundreds of photocopy shops. A fall in electricity consumption would lead to bankruptcies among companies building power plants.

Let us try to imagine what, in a capitalist system, the situation would be once the contraction process has come to an end and a steady state has been reached. The entrepreneurs who survive the contraction process would surely enjoy, by virtue of their profits, a higher standard of living than the average. Since total resource consumption would not be allowed to increase, and total production of goods and services could not increase after the optimum combination of labour- and resource-productivity has been reached, one entrepreneur's sales, in any branch of business, could increase only at the cost of those of another entrepreneur. Of course, there might be attempts on the part of entrepreneurs to form cartels. But the state would try to prevent that. Moreover, there would always be those who would like to become entrepreneurs in order to enjoy a higher standard of living. They would try to push out the existing entrepreneurs. In an SSE, a new entrepreneur's profit, or an existing entrepreneur's higher profit, must lead to another entrepreneur's loss. All may make (higher) profits only in a growing economy. In an SSE, therefore, if capitalism continues, competition would surely become much more brutal than it is today, with many side-effects.

We may be getting a foretaste of these side-effects in the situation in Russia since the early 1990s. There, the economy has contracted – without anybody having taken a decision to that effect – to half the GNP of 1989, and is currently stagnating at that level. Murders of businessmen and bankers by hired killers, policemen waylaying citizens, mafias, widespread corruption and general criminality have all been regular features of everyday life in Russia since the mid-1990s.

Loske *et al.* have dealt with the question as to whether Germany would be able to remain competitive in the world market if, as the only highly industrialised country, it pursued a policy of gradually rising eco-taxes, especially an energy tax. They believe the problem would not be too serious. They think that the process of 'dematerialisation' of the economy would start a process of diminishing international interdependence and stopping the growth of globalisation. Firstly, they argue, 'dematerialisation' would favour the service sector. Then, in Germany, more value would be created through services limited to the region. As a result, there would be an expansion of that part of the economy not subject to global competition. Secondly, they argue, if transport costs rise, traders would not want to get their goods from far-off places.

These arguments are not very convincing. To illustrate the first argument, the authors give the examples of, *inter alia*, repairs and tools leasing. These two activities, of course, create value and in the process also reduce resource consumption. But many of the goods that would be repaired – cameras, tape recorders, computers, and so on – would still be imported from abroad (unless Germans are prepared to pay higher prices for similar products made in Germany). They would have to be paid for through the export of German goods. If the latter become costlier because of eco-taxes, it would be very difficult to sell them. The second argument is also flawed. If transport costs rise in Germany and not globally, then that would be no disincentive against importing goods from, say, Taiwan, for the distance between the port of Hamburg and, say, Cologne is a negligible fraction of the total distance from Taiwan.

But the most important reason why Loske *et al.* think that the competitiveness of the German economy would not suffer, is that they, like Laistner, have a strong faith in the superior power of the brains of German scientists, engineers, and technicians. According to them, '... it is totally wrong to only stare at the differences between production costs in various countries. [What is] decisive is the capacity not to miss the bus in international competition in the area of innovations' (Loske *et al.* 1996: 367). They are convinced that 'the new strengths of the German economy will comprise only to a small degree material products. They will comprise to a greater degree immaterial goods, intellectual property and consultancy services' (ibid: 366).

In plain English, they mean that Germany would sell in the world market more patents, blueprints, etc., and that there would not be much competition in this branch. But how plausible is this assumption? Do the people of the other advanced industrial countries have fewer cells in their brains, or fewer education, training and research facilities? And how much more demand could there be for these superior intellectual, immaterial products in a world that had decided to live in a sustainable manner, for which the necessary technologies and products are already there?

THE STATE AND EQUALITY; ETHICS, MORALS, AND VALUES

Inner contradictions in theory

The authors cited above are not really convinced that eco-capitalism can attain our five goals. Their works are full of inner contradictions as well as contradicting each other.

Let me first point out a few technical contradictions. Loske *et al.* write that 'an ecological structural change is possible *only* in a functioning market economy'. They write that until now the market has been 'hindered' in developing 'itself' ecologically. But they also write that to ecologise it, the market has to be 'domesticated' and that government has to 'steer' it towards an ecological economy by giving it a proper framework (ibid: 169, 175, 367).

Most protagonists of eco-capitalism say that after providing a basic legal–economic framework, the state should refrain from intervention, subsidies, exemptions and the like. Loske *et al.* write:

> There is no economic rationality for giving relief to particular industries. If certain industries are specially hit by ecologically motivated changes in the tax system, then it is those which are no longer sustainable. To ease their tax burden, e.g., the burden of energy tax, with the argument of international competition, is as little sensible as subsidies for dying industries, which have slowed down structural change for a long time. (ibid: 377)

But immediately after this they write: 'For political and social reasons another procedure can be recommended, especially if the temporary effects that the structural change would have on employment are strongly concentrated in certain regions. But then that must be discussed openly' (ibid). As shown earlier, Weizsäcker demands exactly for economic reasons that industries should be exempt from paying ecological energy tax on the energy they need for production. And the Green Party of Germany suggests that energy-intensive industries should be exempt from 80 per cent of ecological energy tax for some years (*Frankfurter Rundschau*, 4 July 1996).

Faced with the certainty that industrial countries that unilaterally raise the prices of energy and other resources would suffer a fall in competitiveness, they demand a customs barrier to protect national industries. Weizsäcker *et al.* and Loske *et al.* demand compensatory import duties: goods that have been produced by highly energy- and/or materials-intensive methods should be charged higher import duties, so that they would not be cheaper than German products.

I do not think this idea can be implemented. Firstly, even if the rules and regulations of the World Trade Organisation (WTO) could be changed or supplemented to make such a policy possible, who could find out how much energy and materials inputs were made to produce the thousands of different products of thousands of factories in a hundred or so foreign countries? If the states of the world sincerely tried to do that, they would have to build another huge bureaucratic apparatus, which contradicts the basic approach of the protagonists of eco-capitalism. Secondly, the WTO cannot compel a state to impose energy and resource taxes at a particular rate. Thirdly, even if that could be done, energy and resource prices and hence also their consumption per unit of production would, for various reasons, still differ from country to country. After all, prices, according to devotees of the market, must not be dictated.

Daly proposes that, in a steady-state US economy, depletion quotas must apply to the sum total of energy and materials produced in the USA and imported from abroad. But he also says that imported finished goods should not be subject to these quotas (Daly 1977: 72). But why? Finished goods are, after all, nothing but transformed energy and materials.

Loske *et al.* want the state not merely to 'influence' the 'innovative capacity' of the economy. They write: 'The most important task that must be tackled *centrally*, that means nationally or, which is the best, at the European level, is the *laying down* of clear reduction *targets* with exact time *plans*' (Loske *et al.* 1996: 369). Jacobs, who speaks of 'constraining' economic activities, says that 'targets need to be set' (Jacobs 1991: 120, 134). But this means that the state lays down targets. They cannot be left to firms or to market forces.

Daly goes a step further. He sees the danger of the concentration of depletion quotas in the hands of a few big monopolies. In order to defend the free market and competition, he demands that the state must see to it that 'no single entity can own more than *x* percent of the quota rights for a given resource or more than *y* percent of the resource owned by the industry of which it is a member' (Daly 1977: 71). Not only that, Daly also wants the state to decide what the 'legitimate economies of scale' are, and to set *x* and *y* so as to allow firms to take advantage of these economies while curtailing monopoly power (ibid). He actually should have used the term 'rationing', which he uses elsewhere in his book.

I agree with such demands. I am convinced that in all industrial countries these state activities – centrally laying down targets, time planning, fixing depletion and emission quotas/permits, determining their rate of periodical reduction, rationing the quotas and permits among enterprises, protecting national industries from foreign competition – are absolutely necessary to attain our goals. But I wonder how much of a market economy would be left among all these state activities?

Of course, establishing monopolies is logically the best path towards contraction and a steady state, because monopolies can avoid the growth dynamic of a competitive market economy. Actually, Daly's proposal would indirectly lead to a quasi-monopoly or cartel. If the firms in a particular industry are each allotted a ration of gradually diminishing quantities of the required resources, then no firm can acquire more, produce more finished goods, and thus push others out of business. Why shouldn't they then come to a price agreement in order to ensure that all that they produce is sold at the highest price that can be extracted from buyers? But the myth of the efficiency of the market and the false belief that human freedom depends on the freedom of the market are so strong that most radical ecologists are unable, or afraid, to come to this logical conclusion. As Andrew McLaughlin argues, 'if the whole economy of a society became one huge monopoly, the imperative of growth could be avoided. But such a society would no longer be capitalist' (McLaughlin 1990: 79).

But it is not necessary to recommend such a drastic step in order to repudiate capitalism. The state activities demanded above are sufficient repudiation. As McLaughlin argues, 'replacing the market as the mode of social decision making amounts to a transition to some form of state capitalism, technocracy, or socialism' (ibid). Elaborating the point, he writes:

> At what point will the political regulation of economic decisions change the system from capitalism into a fundamentally different system? There is a spectrum of possibilities for the regulation of society's interaction with the rest of nature. At one end of the spectrum, the political regulation of most economic decisions would change capitalism into some form of socialism, fascism, or technocracy by eliminating or substantially reducing the role of markets as a mechanism of social choice. At the other end is a pure market capitalism wherein the political regulations of 'private' actions is minimal. There are many possible forms of society between these two extremes. (ibid: 75)

Equality or inequality?

Most of those who advocate a drastic reduction in resource consumption realise that what they are recommending involves a great change. Weizsäcker *et al.* write of the necessity to 'substantially change the technological *and civilizatory* development', and of 'really permanent civilizatory solutions' (Weizsäcker *et al.* 1995: 284, 324). What kind of a society would it be after such solutions and changes have taken place? Much would depend on the degree of equality or inequality. If the rulers were determined to maintain the present degree of inequality in a contracting economy, then authoritarián or even fascistic governance would, I am sure, be necessary.

Some do not see this danger:

> Moderate, or even large income differences seem to be accepted in our society, and differences in consumption power are nothing new. Consequently, the introduction of the concept of environmental space and the increase in prices of products that make use of 'environmental resources' does not change the situation. If only the minimum-wage earners are

assured an acceptable consumption level, the subsequent distribution can be left to the market. Indeed, the rich will then buy up more environmental space than people on an average wage' (Buitenkamp *et al.* 1993: 182)

The authors assume that peace within individual societies would not be endangered by sharp inequalities, but I consider this to be a mistake. In times of continuous economic growth, social peace is no problem, since the lower classes, the unemployed and those on welfare get enough for their livelihood, some even a little more every year. But during a long period of economic contraction, the lower classes would not remain peaceful if inequalities were not ameliorated. Equality is the best means to make necessary cuts in the standard of living acceptable, on a world scale as well as within societies.

Others, however, seem to share my view; reducing inequality is an essential element of their programme and strategy. Meadows *et al.* (1992: xvi) stress the need for 'equity', Weizsäcker *et al.* (1997: 268) speak of 'distributive justice', Loske *et al.* (1995: 32) write of 'social fairness', and Daly even wants upper limits to income and wealth. All propose a guaranteed minimum income for all citizens.

As far as distribution between the North and the South is concerned, there is a clear acceptance of the principle of equality, not just of distributive justice. Even Buitenkamp *et al.* accept this principle. Their definition of the environmental space that a country is entitled to is very strict: 'the world environmental space divided by the world population and multiplied by the number of inhabitants the country has' (Buitenkamp *et al.* 1993: 18). They demand that the North reduce its consumption of resources considerably and help the South to raise its standard of living (ibid). Loske *et al.* and Weizsäcker take a similar position.

It is very strange that the authors cited above are strict egalitarians in respect of distribution between the North and the South, but refuse to apply the same strictness in respect of distribution within their own countries. Buitenkamp *et al.* say in justification of differences in income that that is 'nothing new'. But differences in resource consumption between the North and the South are also nothing new!

It must now be asked whether these laudable intentions and demands can be realised within the framework of capitalism. I do not think they can be – neither within a country nor in North–South relations. The main reason for my negative answer is the utter incompatibility of the fundamental principles of capitalism, namely self-interest and greed, with the ideal of equality or even of restricting inequality. To be a protagonist of any kind of capitalism and to be egalitarian in matters of resources and incomes, is a contradiction. Capitalism, even eco-capitalism, would cease to function if capitalists ceased to be greedy and selfish, and if inequality as a strong motivating factor ceased to exist.

Daly seems to abhor greed. He writes: '... beyond some figure any additions to personal income would represent greed rather than need. ... an income in excess of, let us say, $100,000 per year has no real functional justification, especially when the highly paid jobs are usually already the most interesting and pleasant' (Daly 1977: 70). But why should capitalists invest and risk their money at all unless their profits are much higher than the salary of a highly paid job? Daly forgets that capitalists do not invest and risk their money merely to satisfy their consumption needs; it is their unlimited greed for money, status, and power that motivates them. And this entrepreneurial motivation, without which no kind of capitalism can function, is the functional justification for unlimited inequality.

Moreover, in real life, there are classes and class interests. Capitalists and their associates fight for capitalism with unlimited inequality not because they sincerely believe it to be the most efficient system and the best for the whole society, but because this system best serves their own interests. They would fight for it even if they were convinced that this system is the most inefficient from the point of view of the whole society and nature. For this reason, they even oppose eco-capitalism. Economic contraction, a steady-state economy, equality or even reduction in inequality are, therefore, things that must be put through against the laws of market economy and against the bitter resistance of capitalists.

Doubts about the system

Actually, most of the authors cited here themselves have doubts about the feasibility of attaining their goals within the framework of capitalism. 'The present-day economic system is the central problem. In the whole world, the economic institutions are the strongest force. That is why the necessary changes can be brought about only with them' (Loske *et al.* 1996: 191). Here, it is not the efficiency of the economic institutions that is given as the main reason, but the fact that they are the strongest force (without whose approval nothing can be done). Elsewhere, Loske *et al.* contradict all their previous assertions:

> Whether the system-logic of a market economy is really incompatible with sustainability or whether it must (and can) be overcome, we do not know.

> ... Only if it turns out in future that a reduction in consumption of energy and materials is not compatible with the system-dynamics of a market economy, shall we have to consider other ways of running the economy. Only in that case would society face the choice either to fundamentally change the market economic system or to forgo ecological adaptation in the direction of sustainability. (ibid: 373)

Weizsäcker *et al.* ask 'whether our form of economy can at all tolerate such progress in efficiency as causes it to slim down' (Weizsäcker *et al.* 1995: 173). Their answer, we have seen, is yes. But they also write:

> Global competition is more and more being felt as a torment and threat. Industrialists and politicians are using it as an excuse for ... rescinding already existing social and environmental policies as well as for killing jobs. (They say) ... 'We cannot afford the extra costs. If we demand this and that from our industries that have a negative environmental impact, then they will emigrate, and that would be worse for the environment.' In individual cases, this statement is often correct. But especially if it is correct, then something is wrong with the system. For that would mean that the world cannot afford free trade. (ibid: 309)

Elsewhere, they write of 'war and other conflicts or irrational behaviour under the murderous pressure of world-wide economic competition' (ibid: 296). In yet another place, they inform us of the view of Richard D'Aveni, a protagonist of free trade, who says that (paraphrased by Weizsäcker *et al.*): '... it is no longer a question of competition with the others, it is a question of destroying one's competitors. "Competition is war", writes D'Aveni very clearly. Morals and ethics do not have a place in international competition, he asserts' (ibid: 310). Weizsäcker *et al.* write these things in the context of world trade. But every word of these quotes is also valid for the inland trade of any capitalist country.

After describing the environmental destruction wrought by free trade in the countries of the South, they write in a generalising manner:

> we do not say that somewhere there is a dark conspiracy against the environment. We rather think that the fundamental conviction prevailing in industrial and business circles is mainly responsible for this tragedy, the fundamental belief that in the final analysis the market produces the best result for all concerned. (ibid: 313)

In spite of all this, they remain adherents of capitalism. They want only to 'tame' it, to transform it into eco-social capitalism, 'without setting up protectionistic hurdles' (ibid: 312). They write:

> All these observations should not mislead us into believing that closing the borders would save the environment. In fact, the free market can also help spread eco-friendly methods and technologies. … But something must happen in order to make the principles of free trade really compatible with global protection of the environment. (ibid: 314)

But they are not blind. They know that there could be another solution to these problems. They write with reference to the criticism of free trade: 'The objection that would follow, namely that the socialist planned economic system with its compulsive dissociation from the world markets has been ecologically even worse, is of no avail. For who says that the Soviet type of socialism is the only alternative to total free trade?' (ibid: 309). That is right. Logically, their next sentence should have been: a different kind of socialism, or capitalism without free trade, are possible alternatives. But they do not write that. Instead, they have recourse to modesty. After critically describing how free trade and liberal market-economic policies have caused a relative dismantling of the welfare state, they say in a despairing mood: 'We ourselves do not know yet a real way out. But it appears important to first have the clear diagnosis that free trade, at least in the short run, clearly favours capital and that somewhere, for ethical–political–ecological reasons, there must also be new limits to competition' (ibid: 312).

Capitalism plus co-operation plus ethics?

So finally these protagonists of eco-capitalism arrive at ethics. They want to give the market 'the necessary moral framework' (ibid: 180). Weizsäcker *et al.* are not alone in this. Daly writes referring to his three institutions:

> But … these institutions could be totally ineffective. Depletion quotas could be endlessly raised on the grounds of national defence. … People at the maximum income and wealth limit may succeed in continually raising that limit … extolling the Unlimited Acquisition of Everything as the very foundation of the American Way of Life. … Thus we are brought back to the all-important moral premises. … A physical steady state … absolutely requires moral growth. … Institutional changes are necessary but insufficient. Moral growth is also necessary but insufficient. Both together are necessary and sufficient, but the institutional changes are relatively minor compared to the required change in value. (Daly 1977: 75)

Weakening their pro-market position considerably, Loske *et al.* write:

> It is often argued that in the final analysis also the protection of the natural basis of life can be ensured only through the price mechanism. … That is

> largely true … But can everything and should everything be allowed to be priced and treated in the same way? The draining of a swamp or the building of a motorway cannot be set off against the value of living beings dislodged from there. (Loske *et al.* 1996: 174)

Their conclusion from these reflections is: 'In the beginning there must be a value decision. Only then should we ask in which way the goal that is recognised as right can be best attained. Only here the question of appropriate instruments arise' (ibid: 175).

Quite right. But after reading these passages one wonders how these authors could at the same time be protagonists of eco-*capitalism*. The wonder becomes greater when one finds Daly writing (with Cobb) twenty years later of the 'corrosive effect of individualistic self-interest on the containing moral context of the community' (Daly & Cobb 1990: 50). Obviously, all these authors have a capacity to overlook inner contradictions, of which Daly and Cobb supply the following example: on the one hand, they advocate policies 'that would intensify competition' in the US national market, but, on the other hand, they demand that capital should be attached to regions, and industries should 'identify their well-being with that of states and cities, working with them for the well-being of the community. …' (ibid: 291, 293).

Loske *et al.* espouse co-operation. One thinks of this as a socialist value opposed to the capitalist value of competition. But for Loske *et al.* there is no contradiction between the two. With reference to the postulate of competition, they write:

> Doesn't it contradict the necessity of cooperation as demanded … in our guiding principles? The answer is very unambiguously 'no'. On the contrary … when we speak of competitiveness, we do not mean the ability to push out all the others as far as possible. This would be a very narrow notion of a market economy. It is more a matter of conditions of action and development of the whole society, of all societies. In this matter, cooperation between the economic actors is a constitutive element. … Growing division of labour and interdependence actually demand the most different forms of social cooperation'. (Loske *et al.* 1996: 373–4)

The authors have here mixed up two very different things. Co-operation between firms belonging to different branches of industry is necessary for any production. A firm that builds houses and the firm from which it buys cement have a relationship of co-operation (although both try to exert pressure on the price). But within the building industry, there is competition in the sense that each firm tries to capture a higher share of the market at the cost of other firms. All who take part in a capitalist market as producers or sellers know that they can be pushed out of it.

This is one of the most important features of the real market. It happens as a logical result of competition; it is an unavoidable tendency.

> Competition is what keeps profit at the normal level and resources properly allocated. But competition involves winning and losing, both of which have a tendency to be cumulative. Last year's winners find it easier to be this year's winners. Winners tend to grow and losers disappear. Over time many firms become few firms ...' (Daly & Cobb 1990: 49).

Daly and Cobb, of course, regret it, for it means that the market has a tendency to erode one of its own requirements, namely competition. But they are, unlike Loske *et al.*, consistent. They do not praise co-operation, nor do they see co-operation as an important or constitutive element of the market. On the contrary, they speak of the necessity of trust-busting (ibid).

In their conception of transition to a sustainable Germany, Loske *et al.* also postulate co-operation between the North and the South. They expect enlightened self-interest to motivate firms of the North to co-operate with and help the South, and to desist from trying to make as high a profit as possible (Loske *et al.* 1996: 274). This is wishful thinking and has nothing to do with the reality and conditions of existence of firms in capitalism.

Daly tries to argue that there is no contradiction between private property and moral growth. Private property, he thinks, can be legitimate if there is some distributist institution that keeps inequality of wealth and income within justifiable limits. In his SSE, this institution is there. 'Without some such limits', according to him, 'private property and the whole market economy lose their moral basis' (Daly 1977: 53). Elsewhere Daly speaks of 'ethical boundaries' within which the economy should function (ibid: 69).

I think that upper limits to wealth and income and a guaranteed minimum income make private property tolerable but not legitimate. Daly argues, quoting John Stuart Mill, that 'private property is legitimated as a bastion against exploitation' (ibid). We can add that it is also legitimated as a source of security against unemployment, sickness, and old age. But such a source of security is not needed if there is a guaranteed minimum income and, let us add, institutionalised health insurance. Moreover, would private property in Daly's SSE be only a bastion against exploitation? In the form of industrial and business capital, it would be, even in his SSE, chiefly a means of making more money than the others. That one firm (group of workers) would profit at the expense of another firm (group of workers) poses, strangely, no ethical problem for him.

In order to legitimate private property and give it a moral basis, Daly adopts John Locke's notion that property should be acquired through personal effort. In this notion, an inheritance or windfall lacks the essence of reward for personal effort (ibid: 55).[3] This is very good, and if the upper limit to income is not too high, Daly's SSE could be accepted by socialists. For then, any private property, including bank balance and shares, would cease to be private property after the death of the person through whose personal effort it has come into being. Thus all present-day private property would, in the course of time, become the property of the whole society. This would constitute a rather rapid transition to socialism. Daly's SSE, or any moral economy, would necessarily be or become a socialist SSE. In such an SSE, individuals would be allowed to start and operate a business through personal effort, perhaps jointly with the efforts of others as equal partners (co-operatives), but nobody would be allowed to profit or acquire private property through the work of hired labourers. This was roughly the system of limited private enterprise that existed in the erstwhile 'socialist' society of the USSR.

In the period of transition from Daly's capitalist SSE to a socialist SSE, private ownership of large stocks of means of production would gradually wither away. The entrepreneurs, who would be allowed to start and operate an enterprise with their own effort, would own only a small part of society's total stock of means of production. The greater part of it would belong to society as a whole. In such a situation, basic economic decisions cannot be left to market forces, because the class of capitalists with their motives of self-interest and greed would be rapidly dwindling. Some sort of social decision-making would become not only desirable but also necessary.

Planning would be absolutely necessary in the transition period to ensure an orderly retreat from growing, capitalistic economies. Only through a planned retreat would it be possible to absorb the strong negative impacts of destruction of capital, to prevent mass unemployment, and so to avert the danger of the economy and society collapsing into chaos, war and civil war.

The inner contradictions in the positions and arguments of the protagonists of steady-state capitalism are so obvious, and the logical conclusions I have drawn from their premises so compelling, that I wonder why all these intelligent people are blind to them. I believe they are not blind. They are only afraid of slaughtering the sacred cow that the market economy has become in their societies.

The leading Dutch economist, Johannes B. Opschoor, whose position is close to those of the authors cited in this chapter, admits the dilemma.

He speaks explicitly of the necessity for 'sectoral development and investment *plans* based on government–industry agreements, zoning, infrastructure supply regulations, quota ... etc.' He even speaks of the necessity for 'environmental and economic forecasting and planning' (Opschoor 1991: 20; emphasis added). Indeed, the instrument of tradable pollution and depletion permits/quotas cannot function without forecasting and planning. 'In any case, this would extend the powers of government into areas (e.g., economic planning, pricing policy etc.) from where it is actually withdrawing' (ibid: 21). He then concludes:

> The most profound policy to prevent growth would be that of reducing (world) market insecurity and competition. As this *comes close to the very essence of our economic system* and as faith in the existence of alternatives to that system is dwindling rapidly, one cannot but hope that the environmental crisis can be resolved without having to consider changes as fundamental as these' (ibid: 21–2; emphasis added).

I do not see any chance that Opschoor's hope can be fulfilled. The very essence of capitalism stands in the way of resolving the environmental crisis. Moreover, why shouldn't there be an alternative? If moral growth, ethical behaviour, and co-operation are essential conditions for the supposed success of steady-state or ecological capitalism – conditions I believe to be impossible for any kind of capitalism – then one can say that with moral growth, steady-state socialism can function more easily. After all, moral growth, the ideal of new man, co-operation, and limiting inequality are integral and essential parts of the ideal of socialism. And if only moral degeneration could have been prevented in the USSR, then even 'socialism' there could have corrected its mistakes, removed its deficits, and advanced towards socialism. In any case, the most fundamental crisis of industrial and capitalist societies requires fundamental changes for a solution.

NOTES

1. That cannot be said of the *basic principles* of socialism, despite the existence of criminality in the erstwhile 'socialist' societies.
2. For an explanation of the paradox of thrift, read any good textbook in economics.
3. These are not the words of Locke himself, but those of his interpreter, John McClaughry, whom Daly has quoted.

CHAPTER 6

The Alternative – A 'Third Way', or Eco-Socialism?

MARKET SOCIALISM

Socialists not only have to attain our five goals. They also have to overcome two special deficits of 'socialism': the well-known inefficiency of 'socialist' planned economies, and the deficit in democracy and freedom.

In Yugoslavia, Hungary, Czechoslovakia, and the USSR, 'socialists' did try to overcome these deficits – through what is generally called the 'third way'. Given the practical political constraints they faced, they could not go far enough with their experiments. But many socialists in the West have developed a general theory and broad outline of a third way between capitalism and 'socialism'. It has been called 'market socialism', 'socialist market economy', 'socialism with a human face', and 'democratic socialism'. In the following I shall use the term 'market socialism', which I want to be understood as also covering the contents of the other terms.

For its protagonists, market socialism is not just a practical compromise between the ideal and the reality, nor merely a second best solution. According to some of its theorists, it is the socialism that Marx had in mind. I shall first present the arguments and basic elements of market socialism. My critique of it will follow separately.

Economic democracy, freedom, emancipation

There exist elaborate theoretical arguments for market socialism. Bischoff and Menard (1990) criticise, firstly, the traditional leftist view that socialisation of the means of production is absolutely necessary to ensure that society becomes the conscious planner and executor of its own development, and that, for the same purpose, distribution should be reduced to an act of administration and not left to anonymous market forces. Secondly, they criticise the traditional leftist view that state ownership is the only form of socialisation of the means of production and distribution that makes conscious planning and execution of plans possible.

They assert that there is a fundamental contradiction between state ownership and democracy. If the state owns all means of production and distribution, then it must consequently organise *comprehensive* planning, execution and control; members of society cannot themselves choose their economic activities. Under such conditions, democratisation of the economic sphere is not possible. They quote Marx in support of their view: 'Indeed, it [government] would either be the despotic government of production and administration of distribution, or it would be ... nothing more than a board that maintains the accounts for the collectively (*gemeinsam*) working society' (quoted in Bischoff & Menard 1990: 19). Then they comment: 'Limiting the concept of social property to state property allows only the despotic variant' (ibid).

For Renate Damus, state ownership of the means of production and comprehensive planning were not just results of a theoretical error. They were used by the 'political bureaucracy' as instruments for ensuring their domination. By compelling enterprises to give top priority to fulfilling the plan, the targets of which were determined at the centre, and by preventing them from buying their inputs from each other in a horizontal relationship, the rulers tied the producers to themselves. According to her, such mechanisms are 'fundamentally opposed to decentralisation and democratisation' (Damus 1990: 70).

The concept of social emancipation, including emancipation of labour, plays an important role in the theoretical argument for market socialism. For Bischoff and Menard, 'individual self-activity' should be an important element in a socialist society. Of course, there must be developed social security too, but subaltern status for the majority is not compatible with socialism. It is, according to them, a part of human dignity to be able to participate in and shape all spheres of social and cultural life through one's own efforts and views (Bischoff & Menard 1990: 29–31).

According to Damus, socialist society should fulfil 'the preconditions of far-reaching individual development' in all spheres of life (Damus 1990: 73). But, she writes, 'Freedom, equality, and security do not fit together harmoniously and without conflict. One who decides on freedom and equality as higher values of human existence, cannot guarantee absolute security. It is also not desirable, for, where this kind of thinking gains prevalence, it tends to degrade humans to objects' (ibid: 76). These concepts and ideals are, for market socialists, strong arguments not only for workers' self-management in enterprises but also for allowing members of a socialist society to become entrepreneurs. In this sense, Nove introduces the concept of 'producers' preference', that is, how and in what kind of enterprise people prefer to work (Nove 1983: 129).

> Vertical subordination must, where possible, be replaced by horizontal links, i.e. by negotiated contracts, agreements, with suppliers and customers. This … is a pre-condition both of consumer preference and of producers' preference being satisfied, assuming that customers and producers both wish to have some effective influence on their everyday lives. (ibid: 226)

In other words, according to Nove, not only private enterprise but also market mechanisms are essential for realising these ideals. Damus explicitly demands market mechanisms, 'not only for economic reasons but also for reasons of freedom, for reasons of destatisation.' Then she writes: 'But if one assents to … market mechanisms, … then one must also accept competition and, with it, social insecurity, suffering, unemployment, and inequality among humans. If one does not do that, one condemns people to a one-dimensional way of life secured by the state' (Damus 1990: 76).

Not all market socialists use such forthright language, but all accept the essence of the above quotation. Nove is acutely aware of the contradiction between inequality and socialist ethics. Paraphrasing the Hungarian economist Janos Kornai, and with reference to the need to provide material incentives for the sake of economic efficiency, he writes: '… any system of incentives, if it is to be effective, must be associated with inequalities, and if it is linked with profit [of the enterprise] is bound to conflict with the principle of equal pay for equal work (Nove 1983: 123). A few pages later he writes:

> These species of contradictions are inescapable whenever payment is related to firms' success. Yet if it is not so related, then where is the workers' material interest in the success of 'their' firm? Would they then not be indifferent to the profitability of their firm's operations? The dilemma is an objective fact, and any form of self-management with material incentives will have to live with it. (ibid: 135)

Nove speaks here only of inequality between workers. Inequality would be much greater if one compared the profit-incomes of entrepreneurs with the wages of their hired workers.

Despite her strong words, Damus does think that material security is necessary for free individual development. She opposes only 'absolute security' (she means perhaps job security), which, according to her, would not be necessary in a socialist society, because a 'loss of job would not become a serious problem if people no longer derive their whole identity from paid employment, if informal structures of own-work, neighbourhood relationships etc. are developed, and if people get a guaranteed basic income to protect their human dignity' (Damus 1990: 76).

If private enterprise and competition are accepted, then hiring and firing workers must also be accepted. So there must be a labour market. But, according to Stephan Krüger (1990), the function of the labour market in market socialism would be quite different from that of the labour market in capitalism. It would ensure only the necessary mobility of workers. It would be a social task of the state and enterprises to ensure that workers can always sell their labour power. That would, however, mean that workers must be prepared to acquire several qualifications and move to new places.

Nevertheless, writes Krüger, it would be of great importance to have a system of transferring public funds to private households 'to rectify the differences in money incomes generated by market relations as well as for upkeep payments for non-working parts of the population'. The difference between such transfers in market socialism and social welfare in capitalism would be that whereas the latter are merely expressions of 'class compromise', the former would be expressions of 'desired redistribution'. More decisive, according to Krüger, would be real transfers in the form of 'social–cultural public services'. Such services could also be the beginning of a step-by-step introduction of 'communist principles of distribution' – to each according to his need.

Ownership of the means of production

Private enterprise would be allowed in market socialism. Basing himself on Marx, Krüger writes: 'The character of societies, which generally include several forms of production, is determined by the social relation of production dominating at the time. This applies also to the developed societies of the West' (Krüger 1990: 58). This means that if the greater part of production and distribution takes place through socialist relations of production, then the society can be called socialist even if there are private entrepreneurs and market mechanisms in it.

State enterprises would, of course, dominate the commanding heights of the economy, such as the banking and credit system. But between the two poles there would be various other kinds of ownership: co-operatives, joint ventures between public and private sectors, joint stock companies, companies fully owned by their workers, companies largely owned by their workers, and so on. It seems to me that, for market socialists, all of these would qualify as socialist.

Krüger and Bischoff & Menard quote several passages from Marx to show that, for him, social ownership of the means of production cannot be the same as state ownership. They quote Marx writing that the negation of capitalist production 'does not re-establish *private* property, but it does indeed establish *individual* property on the basis of the achievements of the capitalist era: namely co-operation and the possession in common of the land and the means of production. ...' (Marx 1982: 929).

Ota Sik (1990) brought in the idea of workers' 'capital-participation', through which conflict of interests between capital and labour could be prevented. Workers would not then be guided only by wage-and-job interests. In joint-stock companies, workers should be enabled to become shareholders. But they must purchase the shares, so that the responsibility of an owner is not lost. Sik claims that American enterprises with workers' capital-participation have proved themselves more efficient than those without. Sik thinks that state enterprises should operate only in branches where there is no market competition, such as water supply.

Höltschi & Rockstroh (1985), basing themselves on Ota Sik's work in the 1970s, propose a complete model, which they call 'company of employees' (*Mitarbeitergesellschaft* – MAG). In this model, that part of the profit to be re-invested goes into a fund managed by a kind of trusteeship, which the authors call 'assets administering company' (AAC) (*Vermögensverwaltungsgesellschaft*). Employees also get a share of the profit. But only a part of their share is paid out. The rest goes into the investment fund. It is not stated clearly, but the model seems to expect the original owners or shareholders to hand over their invested capital to the AAC. There are models for the gradual transformation of joint-stock companies into MAGs (Sik 1985: 359). Neither the original owners of the company nor the employees, who become investors, control the capital. The AAC does this. The capital of the company is thus 'neutralised'. This means that neither the original owners nor the workers can sell or retrieve their share of the capital. When a worker leaves the company and joins another, he/she gets a share of the profit of his/her new company. The AAC puts the capital at the disposal of another body, called the enterprise management company (EMC) (*Betriebsführungsgesellschaft*), which actually runs the enterprise.

The company of employees (MAG) is managed democratically. All employees are members of both the AAC and the EMC. The AAC elects its governing body. In the EMC, however, the management is not elected by the employees, and it functions largely independently of them. The employees elect only a supervisory committee (*Aufsichtsrat*), which appoints, controls, and, if necessary, fires the management. Employees organise their work and the production process largely autonomously. Sik spoke of self-managed working groups. The economic sphere is thus democratised, and conflict between labour and capital is overcome. The problem of motivation is also solved: because part of their income will depend upon their own performance, employees can be expected to do their best to make their enterprise function efficiently and remain competitive. By 'neutralising capital', the means of production are freed of their class character. In the course of time, most of the capital of a company will belong to the employees.

Market mechanisms

There is an anti-market socialist position, which holds that socialism and market mechanisms are incompatible, that the latter can be tolerated or used only in the transition period to full socialism, as a sort of concession to underdeveloped real conditions. In reply, market socialists say that this position is based on an untenable identification of the capitalist mode of production with market mechanisms. A Soviet pro-reform economist, Eduard Gorbunow, wrote:

> market, exchange, money, credit are by no means inventions of capitalism; they existed even under conditions of slave and feudal society. Capitalism only gave the market first a national and then also an international dimension. Therefore, existence of market in socialism does not at all mean adoption of any special element of capitalist economy, but only a continuation of the traditions of the world economy. ... (Gorbunow 1989: 59–60)

According to market socialists, the specific mode of production determines the structure and functioning of the market. Therefore, market mechanisms as such do not say anything about the underlying relations of production. That is why, according to market socialists, it is nonsensical to make the market responsible for all the evils of capitalist relations of production, or to think that the cause of subordination of producers to an alien will lies in the commodity form of products of labour.

Bischoff & Menard quote Marx: 'commodity and money are elementary preconditions of capital, but they develop into capital only under certain conditions' (Bischoff & Menard 1990: 107). Elmar Altvater elaborates these conditions as follows:

> Spatial expansion and interconnections are not at all sufficient to transform

> markets into market economy as an economic system if the form of commodity has not been made the general principle of social regulation and if the institution of private property in means of production has not been set up. That means: land must be mobilised as commodity; money, capital, and – above all – labour power must become commodity. Only when the great social transformation has taken place that brings about the breakthrough of this logic, is market economy victorious as a system. (Altvater 1990: 29)

Bischoff & Menard add two more conditions: the separation of workers from the means of production; and investment of capital for the purpose of realising ever higher profits (Bischoff & Menard 1990: 107).

Market socialists not only have no objection to using market mechanisms, they demand their use – for the sake of freedom, democracy, individual development, initiative, efficiency, and so on. Competition must not be abolished, for, according to them, it is necessary as a driving force for innovation and efficiency. They visualise that in a socialist society all the various enterprises, even state enterprises, would compete with each other. Prices must reflect the true costs of production and they must be allowed to be formed through supply and demand. The profit principle must be used, because profit is, according to them, the only reliable indicator that can show that goods and services are being produced to satisfy needs of consumers and no wastage is taking place. But they also want to regulate the market – in general, in order to ensure that socialist relations of production predominate and, in particular, in order to stave off cyclical crises, to ensure growth and full employment, and to ensure that pursuit of self-interest does not conflict with the general interests of society.

Planning

In order to achieve this kind of regulation, all market socialists advocate a combination of plan and market, a 'seamless combination of planned central steering of the national economy with an active role of the market' (Csapo, a Hungarian economist, quoted in Hamel 1972: 180). They reject the Soviet type of planning even if decentralised, because bureaucracy remains bureaucracy with all its negative effects even if power is wielded by regional or local bureaucrats. It would also not help, according to them, if the managers of state enterprises were given more powers within a fully nationalised planned economy. In such a system, even if they had more powers, they would still seek a 'quiet life'. They would, of course, desire more authority, but they would still be officials within a bureaucracy, not entrepreneurs; they would wish only to move up in the hierarchy and enjoy their privileges in security. That would not be consistent with efficient operation, let alone dynamism, of an economy (Nove 1983: 177).

The planning that market socialists visualise is long-term macro-economic planning. It would formulate long-term (15–20 years) goals on the basis of analysis of basic trends of economic development, such as the development of social needs (standard of living, life-style) and of the factors of production (technological changes, sources of energy and other raw materials, population). From these long-term goals would be derived the 'fundamental proportions', such as the proportions of production (of goods and services), the proportion in which the national income would be divided up between accumulation and consumption, the proportion between personal and social consumption.[1] These proportions would determine the particular targets of five-year plans: growth rate of GNP, technological development, investment in individual branches of the economy, state revenues, people's money-income, and so on. But these goals, proportions, and targets would not be binding, they would serve only as 'basic orientation' for enterprises; they would supply information, on the basis of which enterprises would make their own decisions. In short, it would be indicative or framework planning.

But there would be a binding set of economic policies, which would purposefully create conditions 'through which the enterprises would be motivated on the basis of their self-interest to make decisions that would be in consonance with the basic goals and targets of the plan for the national economy' (Kozusnik, a Czech economist, quoted in Hamel 1972: 182). In another formulation, 'the methods of steering would have the effect that through the efforts of the individual enterprises to be successful, simultaneously the centrally determined goals of national economic development would be realised' (Hamel 1972: 180).

The component instruments of such an overall economic policy would cover finance, investment and infrastructure, taxation, credit and interest, wages and prices, and external trade. The central organs would use these instruments flexibly; plan and market would inform and correct each other.

In a democratic socialist set-up, the plan would need democratic legitimation. Höltschi & Rockstroh visualise several alternative plans (drafted by political parties or other organisations), out of which one is chosen by the people through a referendum. Alternatively, the people should vote that party into power whose draft plan convinces them the most.

One very important argument for framework planning and central steering is that this is the only way to prevent the cyclical crises of capitalist market economies. The fundamental proportions, especially that between consumption and investment, if managed flexibly, would ensure that there is neither overproduction nor underproduction.

CRITIQUE OF MARKET SOCIALISM

Let me now present my critique of market socialism. It is not my argument that market socialism violates the teachings of Marx and Engels. The efforts of Bischoff & Menard and Krüger to prove that the masters actually meant market socialism when they spoke of socialism, are futile. Indeed, it seems possible to counter Marx with Marx. For Paul Craig Roberts, Marx considered commodity production and markets as characteristic of capitalism, therefore market socialists are un-Marxist (cf. Nove 1972a: 120). There is also an apparently well-known passage in Engels's *Anti-Dühring*, which Nove paraphrases as follows:

> 'Direct social production' will exclude commodity exchange, therefore also the transformation of products into commodities and into values. The quantity of social labour will also be measured directly, in hours. Production plans will be made in the knowledge of the utility of various products, compared with one another and with the quantity of labour necessary for their production. 'People will decide all this quite simply, without the use of so-called "value".' (ibid: 121–2)

So let us forget Marx and Engels, and examine whether market socialism can help us to attain the five goals set out at the end of chapter 4. But first let us examine whether market socialism is or is not socialism.

Is it socialism?

I stated in chapter 1 that one is a socialist mainly because of the values socialism represents and not simply because of its practical advantages in comparison to capitalism. Already several questions arise.

Bischoff & Menard assert that only market socialism deserves to be called socialism. But they themselves raise a doubt: '... if workers become owners of their enterprise and their enterprises compete with each other for profit, wouldn't that necessarily amount to also socialists now recognising the elbowing mentality as desirable social behaviour?' (Bischoff & Menard 1990: 60). But their answer to this question is vague. Of course, it may be useful for the workers of an enterprise to be its co-owners. They may, for instance, decide that in times of recession nobody would be laid off; instead, everybody would work fewer hours for less pay. But competition between the workers of different enterprises in the same sector remains. Each enterprise would try to capture a higher share of the market at the cost of the others, and the workers of the successful enterprises would get higher rewards at the expense of the workers of the unsuccessful. In times of recession (or zero growth), this competition would be more intense than in periods of growth. What kind of socialism would that be? The worker–owners would then be just like capitalists.

Market socialists accept inequality of income. Nove sees the possibility that in market socialism, even the principle of equal pay for equal work may be violated (socialists have no objection to unequal pay for unequal work). A lorry driver, for example, working in a commercially unsuccessful company would get less than a lorry driver working in a successful one – for equal quantity and quality of work. Commercial success, he writes, 'depends on a variety of causes, many of them quite outside the control of the workers: market fluctuations, imposition of a tariff by a foreign country, and so on. It may indeed have very little to do with effort, or with productivity' (Nove 1983: 134–5). Could the system then still be called socialism? I would say no.

Krüger's idea that differences in money income resulting from the effects of the market should be 'corrected' by means of 'public transfer' of funds to households, may, of course, appear to some leftists to be very socialistic, but it is not. In socialism, one should give to society what one can (from each according to his ability), and get from society goods and services for consumption according to one's work. Public transfer of funds to correct injustices in the economy is not a socialist policy, but charity.

There is also a practical problem here. Such transfers would counteract something that market socialists want, namely the incentive effect of unequal income. Moreover, the differences that would have to be corrected might not be small if market socialism allows private entrepreneurs to make large profits. Of course, the size of private enterprises and their profits can be limited by law, administrative measures or tax rates. Market socialists, however, do not speak of that. And that is understandable; they do not want to kill the spirit of enterprise. But even in a small private company, the entrepreneur would be earning a profit over and above the salary or wage he/she would take as a manager.

Economic democracy

Market socialists have emphasised economic (not just political) democracy as an essential element of the ideal of socialism, claiming that it can be realised more easily in the system conceived by them. Is this a realistic hope?

In a market socialist society, a higher share of the national income may go to the lower strata of society than in capitalist societies. The former would also be democratic in the usual political sense. But since economic power would be unequal because of unequal distribution of wealth and income, the expression 'economic democracy' could not justifiably be used for it. What economic democracy can poor and unemployed people enjoy?

So far as participation in the decision-making process in enterprises is concerned, Nove, paraphrasing André Gorz, rightly remarks: 'The bigger the unit in which he [the worker] is working, the more he is likely to feel alienated, remote from management decisions, a minor cog in a big machine' (Nove 1983: 199). Later, he remarks: '... it is hard to envisage giants of the size of a socialist Du Pont, or Shell, being meaningfully self-managed by the workforce' (ibid: 202). With reference to the experience of self-management in Yugoslavia, he writes, firstly, 'that the desire to participate is by no means universal', and secondly,

> The instrument mechanic, brick-layer, lorry-driver ... will not ... be acquainted with the demand pattern for the product of their firm, or alternative sources of supply of materials, or relative costs of variants, or the economic advantages of this or that innovation. ... So it is not surprising that the Yugoslav experience points to the limitations of participation ...' (ibid: 137)

Referring to the experience of Hungarian producers' co-operatives, he writes: 'few persons other than the manager are concerned with the enterprise as a whole' (ibid).

In Nove's conception, there would be 'a preference for small scale, as a means of *maximising* participation' (ibid: 227; emphasis added). This is surprising. How can the industrial mode of production be efficient with a preference for the small scale? How can labour productivity be increased and other economies of scale be taken advantage of without the large scale? No other market socialists whose works I could obtain have advocated a small scale. And neither they nor Nove have anything against the industrial mode of production, which, in combination with the market, almost compels enterprises to become large.

The limitations of and limits to participation are, of course, there in any kind of industrial society. But the invisibility of the hand of market forces, and the resulting uncertainties, make managing an enterprise more complicated, which, in turn, makes workers' participation more difficult and unlikely.

Could market socialism solve the unemployment problem?

Let us now see whether market socialism is the right framework for attaining our concrete goals.

Market socialists hope that macro-planning and their instruments for steering the economy would overcome cyclical crises and ensure steady growth, so that there would be no big unemployment problem. Nove speaks of 'the duty to provide work' that 'would override considerations of micro profitability' (ibid: 228). But this is again trying to eat the cake and have it too. You cannot base an economic system mainly on the profit motive and then say that the duty to provide work should *override* considerations of micro profitability. Whose duty would it be to provide work for all if enterprises – except perhaps those owned by the workers themselves – were free to lay off workers when they were not needed? In the model of Sik and Höltschi & Rockstroh, even a 'company of employees' (MAG) would shed workers if it should become necessary (Höltschi & Rockstroh 1985: 129–30). Nove himself writes that in Yugoslavia, where enterprises were, at least theoretically, self-managed by workers, there was the 'vexed question of unemployment', which could not be overcome despite a satisfactory growth rate, and even by 'exporting' workers to West Germany. One of the reasons for this was, according to him, that the workers' councils were 'interested in maximising net revenue per worker' (which is what market socialists would like them to do), and had, therefore, '*no material interest* in taking on extra labour'. Hence they 'influenced investment choice in the direction of labour-saving variants' (Nove 1983: 138).

This policy of Yugoslav workers' councils cannot be criticised by market socialists, who, without exception, advocate the use of the most modern technologies to raise labour productivity. They have all criticised the erstwhile 'socialist' economic system because of its tendency to use out-dated, labour-intensive technologies and employ too many workers. In any kind of market economy, competition and the profit motive would compel all enterprises to use more and more labour-saving technologies.

The solution offered by Damus – a guaranteed basic income and some ownwork – would not solve the unemployment problem. It would, of course, be sufficient to prevent or overcome acute poverty. But, as argued at the end of chapter 4, to make society sustainable, it is also necessary that all able-bodied people are meaningfully employed. Moreover, her solution falls far short of guaranteeing social equality and reducing economic inequality to a tolerable level – two other requirements of a sustainable society.

From everybody according to their ability and to everybody according to their work should remain an essential value of socialism. Otherwise it would not be socialism. But that would not be possible in market socialism. For a society that gives priority to this value in times of highly labour-saving technologies must *compel* all kinds of enterprises and institutions to employ more people than necessary and compel workers with jobs to accept less work for less pay (as argued in chapter 5, merely making energy more costly would not serve this purpose). But that would be the end of the market principle, of the autonomy of enterprises, and of Nove's producers' preference.

If equality, solidarity, and co-operation were to become high-ranking values in society, then no compulsion would be necessary. Workers, managers, and their enterprises would, of their own accord, reduce the working hours and incomes of the employed in order to accommodate those who have become unemployed for some unavoidable reason. But that does not agree with the principle of self-interest, which is one of the highest-ranking principles in the theory of market socialism.

The problems created by the world market and the growing globalisation of capital are also intractable for market socialists. Macro planning and steering would be limited to their own states, and that would be insufficient to prevent economic crises and unemployment. Relentless competition at both national and world market levels would compel enterprises to invest more in automation and less in workers. This has already been happening for a long time. Even if all the countries of the world were to have a market socialist economic system, and even if governments could prevent ruinous competition within their own territory, competition would exist between market socialist enterprises of different countries. That would result in crisis and unemployment in some countries and growth plus unemployment in other countries, even if the world economy as a whole grows.

Could market socialism make the economy sustainable?

As shown in chapter 4, economic contraction, at least in highly industrialised countries, is absolutely and urgently necessary. Could market socialism be the right framework for this process?

Most market socialists have not dealt with the question. In most of their writings, I found only unserious passing references to ecology and the need to include environmental protection as a consideration in plans and policies. As to the resource problem, 'Matters are exacerbated by the possibility – or perhaps even the likelihood – of material shortages which obstruct growth in output. To mention this does not imply the acceptance of a full gloom-and-doom scenario à la Club of Rome' (Nove 1983: 5). Elsewhere Nove writes: 'Given that we are discussing an industrialised, developed country, there is no need to assume that *high* growth rates would be a high priority' (ibid: 222; emphasis added). That is, Nove not only fails to question the long-term viability of industrial society, he also expects a modest growth rate to be a part of economic policy in his market socialist state. In fact, with only one exception, all market socialists whose works I have read take it for granted that economic growth would be an important policy objective in their market socialism.

The one exception is Höltschi & Rockstroh. They discuss seriously the radical ecologists' critique of industrialism. But they defend industrial society. They speak of the inadequacy of the concept GNP. But they assert,

> It is not important by how much the economy grows but how it grows. Our goals must be reformulated: How can an economic system achieve optimum satisfaction of the material and immaterial needs of humans? The answer to this question is qualitative and can need a growth of 0 to 5 per cent. Increases in GNP may mean additional protection of environment. (Höltschi & Rockstroh 1985: 129–30)

They refer here to the opportunities of ecological technologies.

Writing in 1985, they reject the view that in highly industrialised countries the material needs of all people have been satisfied. Therefore, in their market socialism there must be a balance between 'growth of prosperity' and protection of the environment. The one should not exclude the other. They believe that their model is more suited to this balance than either the present-day economic system or the 'unrealistic ecological utopias'. Their model, they believe, contains important mechanisms to make the economy more ecological (ibid: 209).

Ota Sik writes that the purpose of his kind of planning would be to plan only the 'macro-economic figures and growth rates', that it could influence 'the tempo and kind of economic growth', and that through it 'the connections between economic and ecological developments could be considered from the very beginning'. He is honest enough to say that 'this would not solve the ecological conflict', but he hopes that 'ecological postulates could be taken over explicitly in the macro plan' (Sik 1985: 360–1). In short, Sik believes it is possible to have economic growth and to protect the environment, even in highly industrialised countries.

Since market socialists do not even ask it, others have to answer the question: can market socialism be the right framework for the necessary process of economic contraction? My answer is: no. Firstly, the inherent growth dynamic of a market economy as described in chapter 5 would obtain also in a socialist market economy and be in contradiction with the necessity to contract. Secondly, there is no reason to believe that the wastes typical of Western capitalist market economies would not take place in a socialist market economy. State or workers' ownership would make no difference, for, in a competitive market, such enterprises would have to behave like a privately owned enterprise. For instance, they would have to spend money on advertisement. Thirdly, the claimed advantages of market mechanisms in respect of efficiency can be realised only in a situation in which both producers and consumers can react flexibly to price signals emanating from the free interplay of supply and demand. If consumers want to buy more of a particular commodity, producers should be able to increase its production. If the price of a commodity is raised, consumers should have the opportunity to shift their preference to an alternative. But in the context of economic contraction, producers and consumers would enjoy little flexibility, so that the claimed advantages of market mechanisms could not be realised.

Market socialists themselves say something similar: '... the market can become fully effective as the mechanism for coordinating the manifold individual preferences only on the basis of high productivity and highly differentiated production' (Bischoff & Menard 1990: 20). That is, on the basis of a highly developed industrialised economy. And: '... the market-type flexibility which will be advocated in the succeeding pages is peculiarly difficult to introduce in a situation of acute shortage' (Nove 1983: 178). Elsewhere Nove writes with reference to some market-like economic reforms in the USSR: '... relaxation of price control and of administrative allocation of inputs will fail to achieve the desired results if there is a sellers' market (excess demand, physical shortage)' (ibid: 125). In future, there will inevitably be physical shortages and sellers' markets. So market socialism, like any other kind of market economy, would be counter-productive, or at least not the right framework, for a policy of economic contraction.

In a situation of shortages resulting from economic contraction, market logic would require that prices be allowed to rise to a level at which there are enough goods in the shops because not many people can buy them. Advocates of market mechanisms would say that this is the only rational thing to do and the people must accept it. But can market *socialists* say that? The question ceases to be a purely economic one and becomes political. If they are really socialists, they would not accept that only the rich should be able to afford certain things, or sufficient quantities of things, and not the masses. After all, socialism has something to do with equality and justice.

This dilemma would not arise for socialists if all people had the same or approximately the same income. But market socialists demand that inequality of income and the possibility of high profits should be there as material incentives. Some even accept unemployment if there is a guaranteed minimum income.

It seems to me that the economy conceived by market socialists should be called simply a mixed economy. Mixed economies are nothing new. India has a mixed economy. Certain things which are called 'social' in a capitalist 'social-market economy' are simply *called* 'socialist' in the conceptions of market socialists. So far as macro planning is concerned, Gorbunow wrote with reference to the actually existing mixed economies of the West:

> Also in the West, there is no pure market economy. ... In the areas of macro-economic, branchwise, and regional planning, the West is in the meantime ahead of the USSR. ... the capitalist system has succeeded in achieving ... an optimal combination of planning-specific and market-specific methods of economic management. (Gorbunow 1989: 60)

Perhaps Gorbunow exaggerates a little. But his point is basically right. This is one of the reasons why we have long heard of the convergence of the two systems. The point I want to make is that neither this nor the other mixed economy would be the right framework for a transition to a sustainable society.

ECO-SOCIALISM

In chapter 5, I have shown why eco-capitalism won't work. Chapters 2 and 3, and the first part of this chapter have shown the kinds of socialism that cannot help us. The only framework that can help us should be called 'eco-socialism'.

Old wine in green bottle, or new wine?

I am not the first to use the term 'eco-socialism'. Many leftists have used it to combine their socialism with their recently developed concern for ecological sustainability of the economy. But the integration is often not convincing. In general, it appears that they still refuse to take the first element of the term seriously. But socialists must not close their eyes to facts.

Let me take as the first example David Pepper's book *Eco-Socialism*, written in 1992: '... the [red–green] project displays potential problems. It tends to accept ... the simplistic ... limits to growth/overpopulation theses. ... However, the socialist contention that there are abundant resources to meet everyone's needs ... has not been convincingly disproved' (Pepper 1993: 247). He dismisses the overpopulation problem as 'old Malthusian (third world) "overpopulation" canards' (ibid: 2). On the resource question, he writes: 'The eco-socialist response to resource questions is ... that there are no ahistorical limits of immediate significance to human growth as *socialist* development' (ibid: 233; emphasis in original). On economic growth, Pepper writes: '... an ecological–communist utopia requires the development of productive forces ... Eco-socialist growth must be a rational, planned development for everyone's equal benefit, which would therefore be ecologically benign (ibid: 219).

It must be mentioned to Pepper's credit that, while speaking of 'everyone's needs', he also reminds us 'that "needs" are to be divorced from our present market-oriented conception of them' (ibid: 247). But this does not appear to make any practical difference, for elsewhere he says:

> It may or may not be true that absolute amounts of copper oxide or petroleum are declining, but this is not relevant ... What we ... consume are telecommunication – and this can now be done by means other than copper wires – and automobile travel – where the 'water'[2]-powered, 'pollutionless' electric car could be a future alternative to the internal combustion engine car. (ibid: 100)

Telecommunication and cars appear here to be both 'market-oriented' and 'eco-socialist' needs.

Enough material is discussed in chapter 4 to see that Pepper is suffering from old illusions. For ordinary citizens of North America and northern and western Europe, it may be difficult to perceive any resource problem at present. But firstly, we are discussing the problems of the whole of humanity, not just those of the first world. Secondly, we are discussing not only matters of immediate significance, but also those of future significance. Anybody who knows the third world knows that there is already a resource and population problem of immediate significance. It is not just that the exploiting rich and middle classes are depriving the poor of their legitimate share of resources. Even in the capitals of India, for instance, the rich and middle classes are suffering from acute shortages of drinking water, electricity, and cooking fuel.

In fact, this problem most leftists have with the ecology and resource crises is quite old. André Gorz recognised the limits to growth as early as 1973, when he wrote that 'pursuit of material growth leads to worldwide dilemmas' (Gorz 1983a: 77–8), and he wrote about the need for recycling, long-lived goods, repairs, and so on. Yet in 1977 he expressed the hope that productive machinery would reach a 'level of technical efficiency where a fraction of the availabe workforce can supply the needs of the entire population'(Gorz 1983b: 147). He made clear whence he expected this rise in efficiency of productive machinery: from microelectronics and automation, 'the micro-electronic revolution' (Gorz 1985: 29). He asserts that 'the technological revolution ... does not just bring down the value of fixed capital per unit of output; it sets up a decline in the *total mass* of fixed capital employed to produce a rapidly increasing volume of commodities' (ibid: 29–30; emphasis in original). He asserts further that modern equipment is characterised by 'a higher performance and a lower cost per unit of output'. And then the climax: according to him, we 'can achieve simultaneously savings in investment ... labour ... and raw materials ... especially energy' (ibid: 29).

Obviously, Gorz believes in miracles. We have seen that a personal computer requires an input of 15–19 tons of materials. The input of materials required by industrial robots, and by automation in general, must be likewise very high. At least the *operation* of personal computers does not require much energy. But industrial robots and other automatic machines require a lot of energy to operate – more than manual equipment. Moreover, seen macro-economically, they lead to an increased total consumption of resources, since the workers they replace do not disappear but continue to live and consume energy and other resources.[3]

Writing in 1988, Martin Ryle seems to take the ecological and resource crises more seriously. But he too seems not to have realised the full depth of the problem. He writes of the 'negative, constraining limits of ecological responsibility'[4] . But he writes of 'eco-contraction' only 'in some branches of industry (automobiles, defence, agro-chemicals)'. He says that 'in resource-intensive areas (energy and transport especially)', his ecological perspective would include 'a commitment to bringing UK levels of resource consumption down to levels compatible with global justice' (Ryle 1988: 66, 76, 100) But he neither defines 'compatible with global justice' as clearly as the Friends of the Earth Netherlands, nor states a concrete figure, such as Schmidt-Bleek's factor 10.

Eco-Marxism?

An effort to develop Marxism in order to reconcile it with an ecological position must also be mentioned here. It begins by pointing out Marxism's shortcomings in respect of the double problem of ecological crisis and limited natural resources. (cf. Deléage 1994), and results in an 'ecological Marxist' theory (O'Connor, J. 1988; 1994).

Deléage criticises Marx for leaving 'the path [he himself] opened by the concept of society–nature totality, which could lead to a fruitful reflection on the interplay of social and natural determinations ...' (Deléage 1994: 47), and for propounding a theory of value that considers labour as its sole source and 'attributes no intrinsic value to natural resources' (ibid: 48). He asks:

> But can a parallel not be established between the first mystification of the economy, the hidden mechanism by which surplus value is formed, and another, unsuspected by Marx, the hidden cost of things subtracted from ecological systems? Should the theoretical status of this concept of ecological cost not be ranked on a par with that of surplus value? (ibid: 48)

Then he concludes: 'Combining the two sets of problems remains one of the most urgent tasks' (ibid: 50).

James O'Connor writes:

> Marx wrote little pertaining to the ways that capital limits itself by impairing its own social and environmental conditions hence increasing the costs and expenses of capital, thereby threatening capital's ability to produce profits, i.e., threatening economic crisis. ... Marx never put two and two together to argue that 'natural barriers' may be capitalistically produced barriers ... In other words, there may exist a contradiction of capitalism which leads to an 'ecological' theory of crisis and social transformation' (O'Connor, J. 1988: 13, 15).

From this, he logically concludes that 'there may be not one but two paths to socialism in late capitalist society' (ibid: 15).

Readers will have noted that my theory is close to these views. But there is also a difference: whereas O'Connor (but not Deléage) speaks of a contradiction between capitalism and ecology, I (and partly Deléage) believe that the contradiction is between any kind of industrial society (today also population growth) and ecology. That O'Connor's position is Marxist but not really ecological becomes clear when, after making capitalism responsible for the present-day ecological and social crises, he writes: 'In these ways, we can safely introduce "scarcity" into the theory of economic crisis in a Marxist, not neo-Malthusian, way' (ibid: 26). At least here it appears that he denies the existence of ecological and resource-related limits to growth, 'natural barriers' independent of the socio-economic system.

What I regret is that neither Deléage nor O'Connor spell out clearly in their essays what kind of socialism they think is possible according to their ecological Marxism. In an essay published in 1994, O'Connor asks: 'Is sustainable capitalism possible?' The answer is, of course, no. But the next question should be: can socialism be sustainable? O'Connor has not said clearly that a drastic economic contraction is necessary if late capitalist (advanced industrial) societies are to take his eco-Marxist path to socialism.

Indeed, such a path to socialism would be un-Marxist. Marx's analysis of capitalism can, of course, be developed and enriched by adding a 'second contradiction of capitalism', resulting from the impossibility of limitless supply of what Marx called 'conditions of production' (O'Connor, J. 1994: 162–6). But a socialism that does not presuppose development of the productive forces up to the level of an advanced industrial economy, would not be Marxian socialism. It is difficult to see how Marxian socialism, so deeply embedded in the growth paradigm, can lend itself to ecologisation. Is this a problem? Much in Marxism is true and valuable, and that would survive and be useful for us in spite of the above conclusion. But it is not our task to *save Marxism*.

Wrong critique of socialism

To the extent that socialists offer one or other variety of old wine in a new green bottle, they are being justifiably criticised by radical ecologists. The latter are not criticising the former's continued adherence to the ideals of equality, justice, classlessness, emancipation, and so on, but the fact that they do not take the ecological and resource crises and their implications for the economy seriously.

But there is a critique of socialism coming from radical ecologists that is wrong. Sandy Irvine is, of course, right in saying that socialism brings with it much 'anti-ecological baggage'. He is also right in deploring the fact that 'traditional left-wing ideas have tended to take hold' among German (and other) Greens. But he is wrong in thinking that a 'promise of a fresh, exciting and relevant vision of a sustainable way of living' lies in a Green politics of 'neither left nor right but in front' (Irvine 1995a: 2, 16).

Ever since most genuine socialists left the party in frustration, the German Greens are indeed practising a politics of 'neither left nor right'. But they are not in front. The Greens have become a party of timorous environmentalists attempting to bring in a few petty environmental reforms, and the majority of them have become adherents of eco-capitalism. Their programmes and policies are full of inner contradictions, which arise from the fact that they are afraid of telling voters hard ecological truths (see Sarkar 1994).

Irvine writes: 'the very nature and vitality of an ecologically informed politics could be at stake in *any* "red–green" fusion' (Irvine 1995a: 2; emphasis added). But why? If socialism's essence can be freed from its anti-ecological baggage, and if socialists learn the true ecological lesson, what then? Then, according to Irvine, they can play a 'positive part', 'be relevant' in creating a more sustainable society in the next century (ibid: 1, 16). But if socialists can only play a positive *part* and be only *relevant*, what is the socio-economic framework that would enable us to create a sustainable society? Any framework other than eco-socialism would be a variety of eco-capitalism, at least in the transition period. But we have seen that eco-capitalism won't function.

It is clear that Irvine's main objection is to a 'fusion' of red and green. I guess he thinks that socialism and ecologism cannot be fused. In contrast, I believe that such a fusion – I prefer to use the term 'synthesis' – is not only possible but also necessary for our common purpose. Certainly, socialism must first learn the ecological lesson. But this is possible. After all, socialism and Marxian (not to speak of Leninist or Stalinist) socialism are not identical. Whereas Marxian socialism, logically, includes the anti-ecological baggage of Marx (and Engels), there is no such narrow affiliation in the case of socialism. William Morris, the utopian and the anarchist socialists, and many others had different conceptions of socialism.

I believe that ecological politics, unless it undergoes a synthesis with the egalitarian, anti-exploitative ideals of socialism, can, in a superficial form, become very much an integral part of the present world economic order, which all sincere ecologists must want to overcome – even if they are not in favour of a fully socialist society but want to attain the limited goal of a sustainable economy. An example of this integration is the 300-hectare *Finca Irlanda* in Chiapas, Mexico. It produces 'biological–dynamic' coffee, for which eco-conscious consumers in Europe and North America pay a higher price, in what is claimed to be 'fair trade'. But in reality, in the harvest seasons, most of the pluckers are brought from Guatemala, because not even poor Mexicans can bear the low wages and hard working conditions. The pluckers, who are paid piece-rates, can earn after a day's work only one-third to one-half of the official Mexican minimum wage (cf. Antillan 1996: 10–11).

It is not only ideals that compel us to accomplish a synthesis of socialism and radical ecologism, but also a concrete, logical necessity, namely that without planning, an *orderly* retreat from today's madness will not be possible at all.

In what has been said so far – from chapter 1 to this page – the theoretical foundation of my kind of eco-socialism has been laid. Before proceeding further, let me sum up the argument contained therein.

- Not only the economies of the world but also the societies must become sustainable. Unsustainable societies – those plagued by war, civil war, social conflicts, chaos, corruption, criminality – cause an acceleration of environmental degradation and resource depletion, which are resulting in any case from our present-day economic activities.
- To achieve sustainability, the industrial economies must contract, with the aim of reaching a steady state.
- This contraction would entail accepting a lower standard of living (but not of happiness) than today.
- A lower standard of living can be accepted by the people if the sacrifices are borne proportionately, following the principle of capacity to sacrifice, as in the case of progressively rising income tax.
- Equality is the best means of achieving people's acceptance of a policy of economic contraction. Equality would also be necessary – both in the period of contraction and in the low-level steady state – for a society's ability (a) to guarantee a certain minimum of goods and services to all, and (b) to prevent escalation of social conflicts.
- The retreat must be planned and orderly. A disorderly retreat would lead to chaos and breakdowns. Planning would have to be comprehensive, with price controls, not just the indicative or framework planning that market socialists suggest.
- In countries with a growing population, the most important and urgent task is to stop population growth, for which state action would be necessary.
- Moral growth, a moral economy and society, are necessary to achieve sustainability. This is not possible in the framework of any kind of capitalism; it is possible, though not guaranteed, in a socialist framework, because socialism is a moral project.

This argumentation for socialism is very different from those ranging from Marx and Engels to Gorbachev. It results from the paradigm shift discussed in chapter 1. It would also be a different kind of socialism – eco-socialism. I am raising the question as to which element in the traditional conception of socialism is more important: developing the forces of production or changing the relations of production, with the goals of equality, a classless society, abolition of exploitation, and emancipation.

We have seen in chapter 4 that development of the productive forces in the usual sense of the term leads to an increase in materials flow, which causes further environmental degradation and depletion of resources. In fact, large parts of today's forces of production must be deactivated if we want to pursue the goal of sustainability. But changing the relations of production is possible. I agree with Otto Ullrich, who writes:

> Socialism is a question of social constitution, of relationships of humans to each other. It is unnecessary ... even fatal to connect this question with an undefinable minimum technological and organisational development of equipment of work. (Ullrich 1979: 21)[5]

Developing the idea further, Ullrich writes:

> There is no *lower* limit of the 'development of productive forces' below which socialism is impossible, but there is an *upper* limit. The level of industrialisation that has been reached today by the FRG and the GDR[6] is creating, via technology, a social structure which by itself makes a socialist relationship between humans impossible. (ibid: 102; emphasis in original)

This means that socialism is possible even in an 'underdeveloped' economy. Ullrich's arguments for his contention in the second quote I shall take up below. Let me first discuss the question of need satisfaction, which is connected to questions of the development of productive forces and resource consumption.

Needs

One might say that a poverty-stricken society does not deserve the attribute socialist. Would a steady-state socialist economy be poverty-stricken? The following equation, which environmentalists often use to show the causes of environmental degradation, is relevant in this regard:

Impact = Population × Affluence × Technology.
(Meadows *et al.* 1992: 100)

It says that if technology and environmental impact are kept constant, affluence/poverty would depend on the size of the population. So a society can itself to a large extent determine its affluence.

What is poverty and what is affluence? Marshall Sahlins, the well-known anthropologist who studied the economy of hunters and gatherers, writes in a paper entitled 'The Original Affluent Society':

> ... there are two possible courses to affluence. Wants may be 'easily satisfied' either by producing much or desiring little. ... There is also a Zen road to

> affluence, departing from premises somewhat different from our own: that human material wants are finite and few, and technical means unchanging but on the whole adequate. Adopting the Zen strategy, a people can enjoy an unparalleled material plenty – with a low standard of living. (Sahlins 1974: 1–2)

Statements like the above can be justifiably criticised as pure romanticism if they are not qualified by the three important conditions of life of the pre-colonial hunters and gatherers, in whose huts the early anthropologists always found material plenty: they lived amid rich hunting, fishing and gathering grounds; their population was stable or growing at a very slow rate; they could (temporarily) migrate to (almost) equally rich grounds, of which there was in those days no shortage (this was also the case with those who lived by slash-and-burn cultivation). Moreover, the material plenty the hunters and gatherers enjoyed was relative to their low or 'underdeveloped' needs. Nevertheless, the point remains valid: if the population remains stable and the material wants or needs do not overshoot the possibilities given by the resource base, then an economy can be in a steady state and the people enjoy a certain degree of material plenty even without much technological 'development'.

It is unnecessary to try to make a sophisticated distinction between wants and needs or between needs, satisfiers and economic goods, as some theorists do (see Max-Neef *et al.* 1990: 31ff.). Whatever one calls them, they must remain within the limits of sustainability. That should not be difficult. For 'the needs that must by all means be satisfied so that a human can survive are few in number and low in level. They are limited to food and loving attention from other people. Already in respect of clothing and housing the range of variations is great' (Ullrich 1979: 102). But the problem with most leftists (even with many who call themselves eco-socialists) is that they cannot accept such an attitude towards needs. Not only for 'the true realm of freedom, the development of human powers as an end in itself' (Marx 1981: 959), but also for their modern conception of emancipation, they think they need absolutely a certain material affluence as the basis. Thus, in 1984, some Green–Alternative leftists in West Germany held a large, high-level congress with the title 'Ecology Between Self-Limitation and Emancipation'. The main question the congress discussed was formulated: 'Is nature so constituted that it allows us a process of emancipation and development, or rather so that it allows, on pain of destruction, only adaptation to its iron laws, so that ecological politics could only be one of imposing limitations and frugality, and dedevelopment of society?' (*Bildungswerk für Demokratie und Umweltschutz* 1984).

The answers one generally gets from eco-conscious leftists and most eco-socialists are vague, ambiguous. On this question also, on the whole, they offer old wine in a new green bottle. For example: '... there are natural limits to every human's material needs. They are needs which can therefore be met within the broad limits of nature's ability to contribute to productive forces' (Pepper 1993: 219). But then he continues:

> The fact that in socialist development people *continuously develop their needs* to more sophisticated levels does not have to infringe this maxim. A society richer in the arts, where people eat more varied and cleverly prepared food, use more artfully constructed technology, are more educated, have more varied leisure pursuits, travel more, have more fulfilling relationships and so on, would likely demand less, rather than more, of Earth's carrying capacity ... (ibid: 219–20; emphasis added)

I am not sure that all this would demand less of the Earth's carrying capacity. Since in Pepper's eco-socialist world not just a minority but all people would develop their needs to such sophisticated levels, the result might be just the opposite. Pepper's list of needs, of course, contains such immaterial things as 'enjoying' arts, eating 'cleverly prepared' food, more education, fulfilling relationships. But it also contains travelling more, which requires higher material inputs. Eating 'more varied' food might mean importing things from all over the world, and 'more varied' leisure pursuits might also mean highly materials-intensive pursuits.

Elsewhere Pepper writes: 'Eco-socialism would change needs, redefining wealth along William Morris's diverse lines, which also include a 'bottom line' of reasonable material wellbeing to all. But all these material needs can be met ... although generally human needs will always become more sophisticated and richer in socialist development' (ibid: 233). The all-important question here is how we understand the concept 'bottom line'. My experience in Europe as well as with the middle class in India gives rise to the fear that it is understood as at least the average standard of living of a central European or North American skilled worker in the 1990s. We have seen that Pepper includes telecommunication and automobile travel. Why shouldn't he, when, according to him, no resource problem exists? In this conception of eco-socialism, it should also be no problem if every citizen (or at least every family) in the world owns and uses a car.

But there is a more basic problem with the concept of a 'bottom line'.

> In a system that tries to satisfy needs through material production (only) ...

> there will always be for every level of 'material wellbeing' some new unfulfilled basic material needs, above all because this system is necessarily very inventive in the production of new luxury goods, which soon become the models of new basic material needs. This system will always be too poor for communism ... What was the day before yesterday the radio, was yesterday the black-and-white TV, is today the colour TV; and tomorrow it will be the 3-dimensional picture projector. (Ullrich 1979: 108)

Of course, Pepper does not say that needs can be satisfied only through material production. But the concept of the bottom line plus the positive value attached to needs always becoming 'more sophisticated and richer in socialist development', necessarily collides with the requirement of sustainability.

Let us take André Gorz as an example. In 1973, his perspective was: 'While *consuming* and working *less* we can live *better*, though differently' (Gorz 1983a: 78). But in 1977 he was talking of 'working less and consuming better'. He called the economic programme of his imaginary ideal President of France 'an alternative pattern of growth'. In this programme, 'every individual will, as a matter of right, be entitled to the satisfaction of his or her needs, regardless of whether or not he or she has a job'; the use of private automobiles would be phased out 'in the most congested urban areas' – presumably it would be allowed outside such areas; travel in cities by trams and buses would be free; there would be hundreds of bicycles at citizens' free disposal. In this programme, nobody must 'feel restrained'; luxuries will not be prohibited; a second residence would cost about 3,000 hours of labour, a private car 600; these labour-costs would 'rapidly decline'. After the successful implementation of the programme, the citizens would have 'earned the right to free work and free time' (Gorz 1983b: 145–9).

It is to be noted that Gorz's utopia was based on the expectation of rising *labour* productivity (not rising resource productivity). In 1983, he wrote that – thanks to the high rate of growth of labour productivity made possible through microelectronics and automation – the citizens of industrial societies would, by the end of the century, need only half of the then (in 1983) usual hours of work per working life to produce everything necessary (Gorz 1985: 41, 116). In this economy, which he saw within our reach, the 'paradise' of the Marxists – 'to each according to his needs' – would finally be realised (ibid: 42); the realm of necessity, which he called 'heteronomous work', would shrink, and the realm of freedom, which he called 'spaces of autonomy', would be enlarged considerably.

In those days many eco-alternatives wanted 'to bring a series of vital activities back to the production sphere of local communities' and to produce for oneself 'important use-values' like food, furniture, repairs, and so on (Bierter 1979: 130–31). But Gorz insisted that in his 'paradise', increased free time would not be used to produce necessary things.

> Reducing heteronomous work does not free time unless everyone is free to use this time as they choose. Necessities must be provided for by another source. Free-time activities, insofar as they are productive, will be concerned with autoproduction of all that is optional, gratuitous, superfluous, of all, in short, which is not necessary, which gives life its spice and value. (Gorz 1985: 57)

The richer and more sophisticated needs, the luxuries, the spaces of autonomy, the increasing free time, in short, the very high 'bottom line' of which Pepper and Gorz speak, may well be desires in the heads of most people in the world, but they are not realistic desires. It is a sort of modern utopian socialism. I do not see any connection between it and a socialist or communist society, even though Marx, the most respected among our teachers, spoke of such a connection. We have more scientific knowledge today than Marx had in the 19th century. Scientific socialism today must accept the fact of limits to growth, must accept the entropy law, or it will not be scientific. There is no use for wishful-thinking socialism.

Some basic issues and eco-socialist positions

I have so far tried to show the compelling logical necessity of my kind of eco-socialism as the only socio-economic framework within which attempts to build sustainable societies in the world could succeed. But how would this eco-socialism look? And why it should look like that?

Marx once said that he did not want to lay down 'the recipes for the canteens of the future'. That was very wise. I do not want to do so either. It is especially difficult to imagine how the world might look when all the non-renewable resources are exhausted. It is also not urgent for us. It will be the task of some future generation. For us, the transition period is more important. The work of building eco-socialist societies must begin today, so we need at least some orientation. In attempting to give this, I shall deal only with some relevant *basic* issues. There is no point in going into details. For an orientation, outlines should suffice.

These issues and my positions on them have two aspects: firstly, they relate to eco-socialism as a long-term *model*; secondly, they relate to policies and practical measures in the transition period – assuming that the majority of people accept the necessity of a transition and that an eco-socialist president or party has come to power. In both cases, my positions are naturally speculative, abstract, theoretical. None the less, they are logically derived from scientific facts. I am afraid the two aspects cannot be dealt with separately, they will get mixed up.

Unemployment, full employment

Even the badly managed 'socialist' economies did not have an unemployment problem. The state guaranteed everybody a job. In *eco*-socialism, however, there would first be a long period of economic contraction and finally a low-level steady-state economy. Wouldn't this create a big unemployment problem? Every eco-activist knows that the issue of jobs is the Achilles' heel of a true ecology movement.

In my kind of eco-socialism, there would be no unemployment problem. Firstly, labour-intensive technologies would be preferred, not only to provide jobs, but also because such technologies reduce resource consumption and, consequently, have less negative environmental impact. Secondly, even in a low-level steady-state economy, there would be a lot of necessary work to be done. Food, clothing, housing, and services like education, health, postal service, and so on, would have to be produced. This would demand much labour, which would be equitably distributed among all who can work. Thirdly, an eco-socialist government would pursue a policy of stabilising and then reducing the population. Eco-socialists cannot but agree with Paul Ehrlich, who wrote addressing leftists: 'Whatever your cause, it is a lost cause unless we control population' (quoted in Weissman 1971: xv). Fourthly, ecologically benign technologies such as repairing, recycling, reusing, manual weeding instead of using pesticides, are all labour-intensive.

Labour-intensive technologies are also socially benign. In capitalist industrial societies, much of the widespread criminality, corruption, social turmoil, violence, and psychic misery are caused directly or indirectly by large-scale unemployment. A socialist government should therefore promote labour-intensive technologies even if there were no ecological and resource-related necessity to do that, simply in order to remove unemployment. In fact, a socialist government should, for this reason, also think seriously about the pros and cons of reducing working hours.

Some psychiatrists speak of the dangers to the mental health of society at large arising from unemployment. Victor E. Frankl warns:

> the unemployed person tends to tell himself, 'I am unemployed, therefore I

> am useless, therefore my life is meaningless'. ... it is not so much the redundancy itself that leads to neurosis as the feeling of meaninglessness. And for checking *that*, also the public assistance system of the welfare state is not enough. ... A human cannot live on unemployment benefits alone ...' (quoted in Nuber 1994: 22)

In 1983, Frankl warned against the reduction of working hours and premature retirement as policies to deal with unemployment. He was convinced that it would generate a feeling of inferiority in addition to the feeling of meaninglessness. He predicted that the feeling of meaninglessness would find expression 'in the form of a mass neurotic syndrome'. Today, many experts speak of a 'depression epidemic' (cf. ibid).

Social security, population control

The quotation from Frankl makes clear that social welfare benefits to prevent unemployment from becoming a private catastrophe is not good for an eco-socialist society. Moreover, neither a contracting nor a low-level steady-state economy would be able to generate so much surplus that society could maintain a large number of non-working, able-bodied grown-up people in addition to providing for the children, the sick, the old, and the infirm. It is also not honourable to demand one's livelihood from society without giving anything in return. That is why a guaranteed basic minimum income without the requirement to work is also not desirable in an eco-socialist society.

A guarantee of some socially useful remunerative work is, for all these reasons, a more appropriate form of social security in an eco-socialist society. Needless to say, raising children or caring for the sick, the old and the infirm, would be considered socially useful work and would be remunerated, even if such work takes place at home and is performed by close relatives. The source of such remunerations could be some special tax paid by those who do not have to perform such duties.

In the third world, during the transition period, social security would have a very strong relevance for a population control policy. It is well known that in most third world countries, children, especially sons, are the main source of old-age security for poor people. In the micro-economic sense, it therefore appears rational to them to have several children, even if it results in a lower standard of living until the children can start working. In order to have two sons in their old age, a couple, on average, have to have five children.

In the transition period, an eco-socialist government would guarantee old-age security to the poor in return for limiting the number of offspring to two. Those who are not poor must pay contributions to an old-age security fund. To induce the poor to accept and practise birth control against their obvious private material interest, this strong material incentive would be necessary. An eco-socialist government could not coerce the poor, and it must in any case solve the problem of old-age security. This policy would also be a means of transferring funds from the rich and well-to-do to the poor – a traditional task of socialists.[7]

The minimum age of marriage would be raised by law to, say, 21, in order to prevent 14–15-year-old girls producing children. With various methods of contraception made available to teenagers, the satisfaction of youthful sexual desire need not be a problem. Feminists especially will find these three elements of policy – guaranteed old-age security, only two (or fewer) children, and marriage at a really adult age – very welcome as contributions to women's emancipation.

Unlike imperialists and the propagandists and ideologues of capitalism, eco-socialists do not believe that population growth is the only cause of poverty and environmental degradation. We know that capitalism, imperialism, exploitation, oppression, and overconsumption of resources by the peoples of the North are major causes of both. But unlike many feminists and traditional leftists, eco-socialists do not think that population growth is a negligible factor. The chief objection of feminists against the hitherto implemented population control policies in capitalist and patriarchal societies cannot be raised against the policies of an eco-socialist government, which would never try to stop population growth at the cost of women's health. It would rather use vasectomy – sterilisation of men – as the principal means for the purpose.

Daly's suggestion of transferable birth licences is conceivable, but only when the transition process is over and the steady state reached. However, even in the transition period, the government would have to tell citizens that the long-term goal would be not just stabilisation but reduction in the size of the population.

Equality, inequality, class differences

We have seen in chapter 3 what great difficulties the Bolsheviks had in implementing their ideal of equality in wages and salaries. But these difficulties arose because of the great emphasis laid on raising production and labour productivity, and because of the policy of rapid industrialisation.

In eco-socialism, the economic policy would be the opposite. I argue in chapter 5 that equality of income would be not only an ideal but also a necessity, both in the period of contraction and in a low-level steady-state economy. A near-perfect equality was also a necessity in the scarcity situation of war communism in the early years of the USSR. The question is whether equality in eco-socialism should mean perfect equality or very much reduced inequality compared to today.

An attempt by some German leftists in the 1980s to implement the ideal of perfect equality is worth mentioning here. The daily newspaper *Die Tageszeitung* (*taz*) paid equal wages, including to editors and managers. All had equal rights in decision making. After about ten years, the policy was abandoned. Two reasons were given for the change: the editors and managers disliked ordinary workers having equal rights in decision making, and they could not really accept the equal, low wages of the *taz*, so that many left as soon as they got a better offer from a more prosperous company, thus creating for *taz* a constant problem of finding journalists (for details see Sarkar 1993a: 263–6). This example is often used to argue that even socialists cannot accept perfect equality, therefore that it cannot work.

The example of *taz* is of limited use in our context. The better journalists, managers, and technicians could get better offers because around *taz* was a highly prosperous, growing economy. The situation in eco-socialism would be the opposite – both in the transition period and in the low-level steady state.

The situation inside *taz* is more relevant to our context. It had a very low circulation, carried hardly any advertising, and its capital base was always precarious. It simply could not pay well. Unemployed or novice journalists accepted the low pay because they had no better opportunity, and *taz* could not pay its workers a lower wage than what was already close to the unemployment benefit rate. The ideal of equality was, of course, a factor, but there was also economic compulsion.

It seems to me that the situation would be similar in an eco-socialist society. It would not have the economic capacity to pay high wages, and it would not be possible to pay wages below a socially acceptable minimum. Nevertheless, there probably would be a policy choice between perfect equality and much-reduced inequality.

Those doing hard or unpleasant work might get higher wages, or work fewer hours for the same wage as others. Another – more idealistic but less practical – rule would be to require everybody to do a certain amount of hard or unpleasant work. People who must unavoidably work longer hours than others might also get proportionally higher wages. But people with high academic qualifications and high intellectual or technical abilities would be in a weaker position. There simply would not be much demand for their qualifications and abilities, as the economy would use mainly labour-intensive, intermediate technologies, and everything would become much simpler. An average Doctor of Philosophy or MSc in physics might have to apply for a clerk's post on the railways, and might be rejected for bad handwriting.

In this connection one must also think of equality between the various regions of the same country (I shall deal with the question of internationalism below). In the USSR and the former Yugoslavia, as shown earlier, the 'socialist' governments did try to bring about this equality. This must also be done in an eco-socialist society. It would, of course, be difficult. For, unlike in the USSR and Yugoslavia, acceptance of such a federal or central policy in the prosperous, well-endowed provinces would not be facilitated by the strong faith in continuously growing prosperity inherent in the growth paradigm. But if moral growth occurs, which is in any case a prerequisite for success in the effort to create a sustainable society, then bringing about equality between the regions would also be possible.

In the speculative situation described above, it is possible and probable that class differences would wither away. Of course, unlike wages, intelligence and natural talents such as those of musicians, artists, poets, mathematicians, inventors, and so on cannot and must not be levelled. They must be encouraged. But possessors of such talents and high intelligence would not have higher incomes because of them. High intelligence would also not be necessary for administering the state or managing the economy, as everything would have become simple (a point I discuss below), so that the danger of the intelligentsia becoming a 'new class' in a future eco-socialist society would not be very high.[8]

So far I have discussed the question of economic and social equality at the macro level. But this might not automatically guarantee equality at the micro level – in interpersonal relationships at home, in marriage, between grown-ups and children, or at the workplace. Within families, in small communities, in religious groups, especially in traditional societies, a wife or a daughter or a sister can be exploited, oppressed, or discriminated against by men – even if she goes out to work and gets equal pay for equal work, and even if she earns the same as her husband or brother. Similarly, children can be exploited and oppressed by grown-ups in the family. Macro-level sustainability of a society might not guarantee that such micro-level phenomena would vanish automatically. Laws and organs of the state may not be sufficient to abolish such evils. That would be a cultural task for socialists, to be fulfilled through social movements.

The motivation problem

The problems the 'socialist' economies (except Yugoslavia) had because of workers' and managers' lack of motivation to work sincerely give rise to the question of how an eco-socialist society would ensure the opposite.

Since eco-socialism would not be market socialism, the threat of losing one's job, self-interest, competition, the risk of making losses or going bankrupt, would be ruled out as sources of motivation. The material incentives used in the USSR are also ruled out, because in a contracting or low-level steady-state economy, more can be paid to some workers only by paying less to others. That would be the end of socialism. Revolutionary enthusiasm and appeals to revolutionary consciousness, which played an important role in the USSR, at least in the first three decades, also cannot be there, because, firstly, eco-socialism would probably not come through a burst of revolution, and secondly, because it would not try to build up a prosperous economy. On the contrary, it would gradually dismantle a prosperous economy, at least in the first world. The majority would accept this as a necessity but would most probably not be enthused. Coercion, also practised in the USSR, would be an antithesis of socialism.

What then remains? Even a low-level, steady-state eco-socialist economy would not function without sincere work. It seems to me that apart from moral growth, without which the process cannot even begin, there is another hope. Since economic activities in eco-socialism would, to a large extent, be decentralised and hence more surveyable, production units small, and local communities largely self-provisioning and themselves in charge (I shall come back to this point below), the welfare of a local community would *visibly* depend on sincere work by everybody, and there would be a real, objective interest to work sincerely.

Of course, there might still be lazy or corrupt workers, but they would be easily seen as such, and there would be social control. The community would shame such people. At the same time, the community would respect people who work for it without demanding money. In such a context, sincere work would become the normal expectation and could be ascertained even without counting the units produced by an individual worker. Piece-rates could then be dispensed with; 'from each according to his ability' could become reality.

I think the Marxist hope that in communism work would become 'life's prime want' (Marx & Engels 1976: 19) was not too absurd. One can observe how miserable a person feels who has no job, although he or she might get enough welfare benefits. One can observe how happy many unemployed or retired people are to get some unpaid but meaningful work in associations and charities. And we can observe how gladly many of us carry waste paper and empty bottles to containers for recycling. We can observe this in capitalist societies today. So why should it not be reasonable to expect that people would work all the more sincerely in an eco-socialist society, in which they would be working for themselves and their community, not for capitalists.

But still, a lot of consciousness-raising work and good examples would be necessary, especially from state and community leaders. But the leaders must not forget, especially in the context of labour-intensive technologies, that people's capacity to work – manually and mentally – is limited. The best policy might be to reduce prosperity and work. The hunters and gatherers Sahlins wrote about worked only three or four hours a day (cf. Sahlins 1974: chapter 1).

Economy, state, planning

I have argued that a contracting economy would need planning. Plans for contraction, unlike plans for growth, can, at the beginning, be implemented only by a strong state. Such plans, even if democratically legitimised and supported by social movements, must be put through against the strong opposition of those who would have much to lose.

In the transition period, the most important question would be what to do with enterprises that close down or whose production is reduced. There can be no question of paying compensation, especially in a contracting economy. But the state can give the owners some kind of pension, or jobs in the remaining economy. Reducing working hours or job-sharing might be necessary for managers as well as for workers.

For conventional economists, the contraction process would be similar to an ever-worsening recession, and the low-level steady state a great crisis without end. There can be no doubt that in such a situation the whole economy would have to be socialised, beginning with nationalisations.

Let us try to imagine what, in such a situation, would happen if enterprises were not nationalised. Because of planning, enterprises not closed down would, of course, be able to sell whatever they are allowed to produce, and the planners might set such prices that the revenues of the enterprises exceeded their costs, that is, they would make some profit. But the volume of profit and also the rate of profit – profit as a percentage of invested capital – would continuously sink until the steady state is reached. Capital in the financial sense would be destroyed to a large extent – not only the capital invested in enterprises that close down, but also much of that invested in enterprises allowed to function with reduced production. Not only would the prices of the latter's shares fall drastically, but such shares would hardly find buyers. In such a situation, no private person or firm would invest in any enterprise. Moreover, it would be unjust if some enterprises were closed down without compensation, while others were allowed to make a profit. Nationalisation of the whole economy would be the only fair solution.

Although capital in the financial sense would be destroyed, the infrastructures, plant, machines, mines, forests, and agricultural land would still be there. They would still be able to produce goods and services. Since their full capacity would not be required, they would be used alternately. Some of them, especially machines, would be stored for later use.

Once the contraction process has been completed and the steady state reached, the tasks of planning and managing the economy would become much simpler – if only because the volume of production and the diversity of goods would go down by a factor of ten (in the first world). Centralised planning with its concomitant problems can then be limited to a minimum.

With reference to his description of the 'snags and disadvantages of regionalization as a solution to the problems posed by the impossibility of centralization', Nove writes:

> If a reader feels that such conclusions are disproved by Chinese experience, the answer is that China's decentralization is based upon a much less developed economy, in which resources within a province are used mainly to satisfy the requirements of that province. ... However, in a modern industrial economy ... local industries producing only for the local market with locally produced inputs are a minor segment of the whole. In China too, the output and allocation of products of large-scale industry of national significance are centrally planned. (Nove 1982: 76–7)

Although a low-level, steady-state planned economy would be very different from the Chinese economy of the 1970s, the point is clear. Decentralised planning and management of the economy would be possible and advisable. Such decentralisation can go beyond the provincial level. The province may be too big. Planning, management, and control at the regional and local community levels are conceivable and would be necessary if a large degree of regional or local self-provisioning is one of the aims. Participation in the planning and management process of those concerned and affected would then be possible, albeit not above these levels. This would definitely increase the efficiency of a planned economy.

Ownership of the means of production can also be decentralised. Regional and local authorities can function as formal owners. If some private enterprise is allowed, the state can lease out or even sell some means of production (but not land) to private entrepreneurs. Property other than means of production can be left in the owners' possession at the beginning of the transition period. But, following Locke's and Daly's argument that property is legitimate only if acquired through personal effort (cf. Daly 1977: 55), such property would pass to society or the state after the owner's death.

Private enterprise?

Since we desire a socialist society not only because it is necessary or has some practical advantages but also and mainly because of the values it represents, it is clear that in an eco-socialist society exploitation of humans by humans and competition among working people for profit or higher income cannot be allowed in any case. But one need not be dogmatic. Subject to these two essential conditions, it is conceivable that, for practical reasons, an eco-socialist society would allow some private enterprise in some areas.

Although decentralised planning would function better than centralised planning, it would still be difficult to plan, direct and control the activities of a large number of small economic units, such as restaurants, taxis, cycle-rickshaws, retail shops, tailors' and other craftsmen's workshops.

An eco-socialist government might allow private enterprises based on a person's own labour and that of his or her spouse and grown-up children. They would not be allowed to hire labourers. A limited number of persons might be allowed to form a genuine co-operative, in which all members have an equal share and equal rights and duties. Since raw materials, intermediate products, and equipment would be allotted to them by the planning authorities, and especially because the economy as a whole would be a low-level steady-state one, there would be little possibility for any such entrepreneur to become rich. At the most, he or she might earn somewhat more than a comparable worker in a state or collective enterprise because he or she was working harder. Moreover, because no such enterprise would be able to expand, higher demand would only attract new entrepreneurs. A fall in demand would induce some closures. In fact, similar possibilities existed in the centrally planned economy of the USSR, and they proved their merit.

Since agriculture is vital for survival, it must be planned. The items the people need most must be produced on a priority basis in the required quantities. Possibilities of bad harvests, and hence reserves, must be included in the planning. But should the production units be collective farms, co-operatives, or small family farms?

Collective farming, much maligned because of the negative Soviet experience, has several obvious intrinsic advantages (economies of scale), some of which could not be realised in the USSR for extraneous reasons. But with the assumed moral growth, with decentralised instead of centralised planning, with much regional self-provisioning, and with stabilisation of population, the negative aspects of Soviet collective farming could be avoided in collective farming in an eco-socialist society.

The hard life of family farmers in capitalism – no holidays, 12-hour days, compulsory early rising every day to milk the cows, male farmers' difficulties in finding a bride, uncertainty about who will do the daily work if the farmer falls ill, the vagaries of weather, uncertainty about the harvest – all these things are arguments for socialist and collective farming. In macro-economic and ecological terms, what is the advantage in each family farmer owning machines and other capacities, which remain unutilised for much of the time? What is the advantage in each farmer travelling to the (wholesale) market to sell produce, a regular feature of third world agriculture which is being increasingly observed in the first world?

In the transition period, and more so in the low-level steady state, the quantity of machines and equipment available to agriculture as a whole would also go down, so that several family farmers (on leased land), if that were the future system, would have to share each item. They would have to buy them jointly or hire them, when needed, from some state or community organisation. In either case, they would have to work out a time plan for using them. All of this is not very far from collective farming.

But an eco-socialist society, if it opts for collective farming, must also think of the diseconomies of very large units of the Soviet type and the advantages of more easily manageable and surveyable smaller units, particularly in respect of motivation and social control. The optimum size would have to be found through research and experimentation.

I have heard the view that when the transition has been completed, the greater part of a low-level steady-state economy might consist of family farms and private handicraft enterprises, or co-operatives that would not exploit hired labour, and that under such conditions it could and should be a market economy. I do not think it would work. All the arguments against a market economy and for a planned and controlled economy would be valid, even under such conditions. At the most, one could visualise a mixed economy with a heavy stress on planning and control, but not a market economy.

Distribution, market, rationing

Market socialists think that the market is indispensable, especially in the sphere of distribution. For example:

> not all economic processes can be planned, and the market remains indispensable within a complex, democratic system of regulation. Moreover, in a society based upon division of labour, people are not only producers: they are also consumers who would be hopelessly overstretched if they had to articulate everyday consumption decisions in a political act. Without market coordination, then, a non-capitalist society based upon division of labour would be unable to function. (Altvater 1993: 254–5).

According to Altvater, 'the association of free humans' suggested by Marx, if it is a 'mass society', cannot work without the help of the market (ibid: 255).

In my view, Altvater's arguments do not apply to a low-level, steady-state eco-socialist society. (1) Such a society would not be a mass society. (2) The division of labour in it would be much less than in present-day industrial societies. (3) It would, of course, be democratic, but not too complex. (4) The quantity, kinds, and brands of products that would be available to consumers would be very limited compared to what is available today in, say, Germany. His arguments are valid as a description of the difficulties that an eco-socialist government might have at the beginning of a transition period. But as the contraction proceeds, the difficulties would become less and less acute.

Because of the low quantity of goods available for distribution, some sort of rationing might be necessary in order to ensure that all who need a product get an equitable or fair share of it at a reasonable, fixed price and do not have to resort to the black market. The black market in consumer goods flourished in the USSR because the government refused to introduce strict rationing in a situation of scarcity. Even capitalist Britain introduced rationing during the Second World War. 'The rationing … constituted a … state intervention (which had a strong egalitarian aspect) in pursuit of an acknowledged collective interest. Of course this was not 'popular' … but despite some evasion it was generally accepted' (Ryle 1988: 62).

In an eco-socialist society, the culture of long-lived products, reusing, and repairing would help to ease the difficulties caused by low production. Sharing consumer durables and tools would also help. This would necessitate the promotion of a collectivist/socialist spirit, to prevent petty quarrels among those sharing scarce goods.

An eco-socialist society would no doubt be unable to fulfil all consumers' wishes (this is not possible even in highly industrialised societies). But many wishes can be ascertained through research, and the most popular could be fulfilled. In respect of clothing, the consumer would be in a better position in an eco-socialist society than in an industrial one; since labour-intensive tailoring technology would be used, clothes would be custom-tailored, rather than factory-made, as is the case even today in underdeveloped countries, such as India.

Regionalism, international trade, globalisation

I have said above that in eco-socialism, to a large extent, economic activities would be decentralised, economic units small, and regions and local communities self-provisioning and autonomous. Eco-socialists need no romantic or parochial love for the local or the regional to pursue such a policy (I shall take up the question of the size of economic units below). While the various components of a cup of 'German' yoghurt travel 8,000 km before reaching the consumer (Böge & Holzapfel 1994), Indian villagers eat yoghurt made 100 per cent in their own village, and ecological impact and resource consumption are very much less. Labour productivity in Indian village yoghurt production is, of course, very much lower than in the German yoghurt industry. But for eco-socialists, increasing resource productivity has the highest priority.

This is our main reason for propounding the general principle: as far as possible, produce locally or regionally. 'Region' can be understood flexibly from case to case. But it should not be misunderstood as a state. Trade between Vladivostok and Moscow is usually considered to be internal trade, whereas that between Munich and Zurich is called external trade. For ecologists, this is nonsense. We should speak of long-distance and short-distance trade. In eco-socialism long-distance trade must also contract. But some long-distance trade would be necessary even in an eco-socialist, steady-state economy, unless we want to become fanatics. Salt, for example, cannot be produced everywhere in India. North Indians must import it from the distant coastal regions. Autarky, self-reliance, semi-autarky and so on have to be considered in relation to needs. The higher the quantity and complexity of goods required to satisfy needs, the greater the necessity for long-distance trade. Our needs and the required goods would have to be kept under control, so that long-distance trade can be kept within sustainable limits worldwide.

So, is the law of comparative advantage wrong? Not as long as its underlying assumptions – relatively low international mobility of labour and capital – remain valid, and as long as transport costs are negligible owing to cheap energy (oil) and low-paid third world sailors. But international mobility of capital is increasing, causing great uncertainty and sometimes havoc on the employment front. (The chaos and crisis generated in 1997–98 in South-East Asian economies by speculative movements of capital, especially of portfolio investments, are another story.) This cannot be to the advantage of all participating countries. Moreover, for ecological reasons energy prices must be raised in order to reduce consumption. And nobody should support exploitative wages for third world sailors.

There is another reason why eco-socialists would generally oppose economic globalisation and, in certain circumstances, aim at a sharp contraction of even short-distance international trade. The economy must be surveyable, predictable and controllable if a government is to succeed in implementing eco-socialist policies in the transition period. But the present extent of international trade and globalisation of capital has already produced the opposite situation. Moreover, transnational corporations and big foreign powers, including the IMF and the World Bank, should not be in a position to dictate policy to weak or small states. Too much dependence on international trade and finance has already robbed such states of much of their independence.

For the above wisdom one can find support even in the writings of Keynes, who, without any knowledge of the ecological problem, demanded 'certain elements of national self-sufficiency' and protection of the home market in certain circumstances (quoted in Moggridge 1976: 104). He also wrote:

> I sympathise, therefore ... with those who would minimize, rather than with those who would maximize, economic entanglement between nations. Ideas, knowledge, art, hospitality, travel – these are the things which should of their nature be international. But let goods be homespun wherever it is reasonably and conveniently possible; and, above all, let finance be primarily national. (quoted in Daly & Cobb 1990: 209).

Third world, first world, internationalism

One may think that the problems of third world economies are different because they are underdeveloped, that whereas first world economies must contract, those of the third world should be developed to some extent. Some third world activists also speak of a right to development (e.g., Agarwal & Naraian 1992: 142). But that is a wrong position. Large parts of the third world are already industrialised and have already crossed the boundary of sustainability, because any industrial economy is, as argued earlier, unsustainable. There may still be some areas in the third world which can be developed without violating the principle of sustainability, and perhaps third world economies in general need not contract to the same extent and at the same tempo as those of the first world. But the difference is only a matter of degree.

Nevertheless, first world eco-socialist governments could and should help the third world – not to become developed, but to accomplish the transition to sustainability. This is partly because nobody who is not an internationalist deserves the attribute socialist, but also because, in the long run, sustainable societies in Europe and North America would have great trouble if the majority of societies in the world were to collapse due to unsustainability (or for any other reason). They would perhaps be overrun by 'armies' of refugees and criminals.

They could help, firstly, by cancelling the external debts of the third world countries, so removing the pressure to produce and export more, and, secondly, by refusing to import sustainability at the cost of unsustainability in the third world (see chapter 5).

In this connection, it is worth recalling that in the 1970s several leftist political economists recommended that third world countries should pursue a policy of dissociation from the world market (cf. Senghaas 1979; Amin 1977). To some this may appear similar to what I am saying. But there is a great difference between their dissociation and an eco-socialist contraction. The visualised purpose of their dissociation was 'auto-centric development' (that is, industrialisation), with the aim of later taking part in the world market as a strong and equal partner. North Korea and China were often mentioned as examples. As we know, North Korea failed to attain this end, and China gave up the policy. The reason for its failure is simple: the growing quantity and complexity of needs and the consequent appetite for capital and know-how requires increased participation in the world market. The eco-socialist economic policy would not have this difficulty, for the government would consciously pursue a policy of reducing and simplifying needs.

If in some distant future the world becomes *one* eco-socialist world, the minimal long-distance trade between its various regions that would be necessary and possible within the limits of sustainability would be just, fair exchange. Perhaps it would then be conceivable to form a world economic council that would plan such trade. One of the objectives of such planning would be to distribute the world's resources equitably, transferring resources from richly endowed regions (oil from

Saudi Arabia, wood from Canada) to poorly endowed regions. But this is too speculative for the present.

Small scale, large scale, domination, democracy

But why, one may ask, must the economic units be small? Economies of scale (despite some diseconomies) cannot, after all, be denied by eco-socialists.

Eco-socialists need feel no romantic love for the small. Small may or may not be beautiful, but it is a necessity and a necessary consequence of other necessary policies. The units producing yoghurt in each village cannot but be small. Units producing for local, regional, or provincial needs would be much smaller than if supplying the world market.

Economies of scale are the main argument for industrial technologies with their high capacities and high prices. Such technologies, once acquired, must be utilised almost fully if they are not to cause loss or waste. They must produce on a large scale for a large market. Only then are they more economical than other technologies. But, for reasons already stated, eco-socialists would opt for labour-intensive, cheap intermediate technologies with low capacity, so that large-scale production by any one unit would bring no advantage and small-scale production would entail no disadvantage.

Eco-socialists would opt for small units for another reason. In industrial societies, the old socialist ideal of workers' control in enterprises has hitherto proved impossible because, generally, production takes place in large units and on a large scale. This kind of production is impossible without the subordination of workers to a higher authority. That was Engels' view, though he surely shared Marx's ideal of society as a 'free association of solidary individuals'. In an article entitled 'On Authority', Engels spoke in this connection of 'a veritable despotism *independent of all social organisations*', of 'a certain authority ...' and a certain subordination ... *imposed* upon us together with the material conditions under which we produce and make products circulate'. He knew that development of large-scale industry 'increasingly tends to enlarge the scope of authority' (Engels 1972: 100–104; emphasis added). We have seen that modern authors like Gorz and Nove hold views similar to those of Engels.

Engels' views were limited to authority in large-scale industry, but Ullrich thinks they apply generally to all institutions of society (the views expressed in this paragraph are based on Ullrich 1979). According to him, anything very big is not easily surveyable, hence it cannot be administered directly by the people involved. It escapes their control and supervision. Real democratisation of *society* is not possible in such a situation. The bigger economic units become and the further the division of labour progresses, the more the workers become mere appendages of ever more sophisticated machines. Such a production system cannot be run as the 'domination-free administration of things' – the hope expressed by Saint-Simon and cherished by generations of Marxists and other communists. Such a production system cannot but be hierarchical, authoritarian and centralised. Hierarchies, moreover, have their own dynamics. They reproduce and reinforce themselves and do not remain confined to industry. 'Such hierarchies, as soon as they cross a certain complexity and size, become necessarily transformed into "social structure".' Ullrich calls this process crossing the 'social-critical limit', which he considers to be a general phenomenon observable in every sphere. Basing himself on the formulations of Marx and other socialists, Ullrich writes that, ideally, a socialist society would be 'a domination-free association of solidary individuals'.[9] This ideal may never fully become reality. Nevertheless, socialists must always try to approach as close to it as possible.

However, a significant difference emerges here between eco-socialism in the transition period and the model of a steady-state eco-socialist society. In the former, a strong state would be necessary to ensure a planned and orderly retreat. But a strong state, even when democratically constituted, does not quite agree with the *ideal* of democracy, which includes democracy even in the economic sphere, and popular participation in all spheres of politics. This contradiction must be tolerated in the transition period.

If the ecological and resource-related compulsions lead to the creation of an economy based on small-scale production with simple, easily surveyable technologies and organisations, then perhaps a close approach to the ideal might become possible. Perhaps, in some distant future, Lenin's ideal that 'any lady cook should be able to run the state' (quoted in Sinyavsky 1990: 151) would not sound absurd – if we had correspondingly small, simple political units, which might not resemble the states of today.

But let us return from the distant to the near future. Workers should control enterprises as far as possible in the transition period. If workers take part in shaping the details of the contraction process in their own companies, then the process in the whole economy would tend to be peaceful and less difficult.

But can we expect workers to take part in closing down their enterprise or reducing its production? I think we can. Let me quote from a report on the ecology movement in Eastern Europe:

> One of the pillars of the ecology movement were the workers. Visitors from the West were again and again surprised to see that the most committed activists in the struggle against environmental pollution were workers of the guilty factories. For example, the workers of a chemical factory in Krasnoperekops, Ukraine, demanded the closure of a part of their own factory. They had no fear of becoming unemployed. After all, the state had to give them new jobs. But that has now changed fundamentally. With the growth of a market economy grew the fear of losing one's job. Today, people would not think of demonstrating against their own boss. (*Frankfurter Rundschau*, 6 April 1993)

In eco-socialism, nobody would fear becoming unemployed.

Eco-dictatorship?

Of course, we hope for a higher level of democracy, but it cannot be denied that there is also the danger of dictatorship. Since contracting the economy would not really be popular, and would perhaps be seen by only a narrow majority to be a necessity, there would be strong resistance to the process, very much more than can be observed today against some proposals for environmental protection. Opponents of the policy of contraction might capture power by undemocratic means and try to continue the policy of economic growth with the help of the army, or fascist forces, and propaganda.

Unlike many who fear that eco-radicals might demand the installation of a dictatorship with the purpose of protecting the environment, I do not think eco-radicals can be so stupid as not to realise that their purpose can be achieved only with the support of the majority. The army might give additional support to the project, but would not be decisive. If millions of citizens are determined to ruin the environment and exhaust resources for immediate gains, then not even the army would be able to prevent it. The majority of the soldiers and officers would probably do the same. But if democratic transition to a sustainable society fails to take place, then, in the chaos and breakdown that would inevitably come in the wake of ecological and resource crises, dictatorships would be set up – although to what purpose is difficult to guess today.

BACK TO GANDHISM?

It may appear that my kind of eco-socialism is similar to Gandhism. This is largely true. Gandhi was a critic of industrial society. He wrote in 1940: 'Nehru wants industrialization, because he thinks that if it is socialized, it would be free from the evils of capitalism. My own view is that the evils are inherent in industrialism, and no amount of socialization can eradicate them' (quoted in Madras Group 1983: 8). Beginning with a dogmatic, total rejection of machinery (Gandhi 1982), he later took a position of cautious acceptance of some industries. But he insisted that in his 'social order of the future', 'the order of dependence will be reversed. Hitherto, the industrialization has been so planned as to destroy the villages and village-crafts. In the state of the future, it will subserve the villages and their crafts' (Gandhi 1966: 122). On the question of cities, he wrote: '… nothing will be allowed to be produced by the cities, that can be equally well produced by the villages. The proper function of cities is to serve as clearing houses for village products' (Madras Group 1983: 7). Gandhi also intuitively understood the limits to growth. He wrote in 1928: 'The economic imperialism of a single tiny island kingdom (England) is today keeping the world in chains. If an entire nation of 300 million took to similar economic exploitation, it would strip the world bare like locusts' (quoted in Bandyopadhyay & Shiva 1989: 39). On the question of ownership of the means of production, Gandhi sometimes took a socialist position. In 1924, when he was told that his beloved sewing machine must also be produced in factories, he wrote: 'I am socialist enough to say that such factories should be nationalized, or state-controlled. They ought only to be working under the most attractive and ideal conditions, not for profit, but for the benefit of humanity' (quoted in Sen 1976: 3). And in 1947, he wrote: 'In the non-violent order of the future, the land would belong to the state, for had it not been said "*sabhi bhumi Gopalki*" (all land belongs to God)?' (quoted in ibid).

In spite of these and other similarities, eco-socialism cannot mean simply a return to Gandhism. Gandhi was not, nor claimed to be, consistent. In his long life, he expressed many contradictory views. Moreover, some of his socio-economic positions were naive, confused, obscurantist, and unsocialist. He did not consider the Hindu caste system to be harmful. On the contrary, he thought it was 'in its origin … a wholesome custom and promoted national well-being'. He thought caste 'had saved Hinduism from disintegration'. According to him, 'caste does not connote superiority or inferiority. It simply recognizes different outlooks and corresponding modes of life' (Gandhi 1965: 1–8). He also made light of the evil of class divisions in society. He wrote:

> Man being a social being has to devise some method of social organization. We in India have evolved caste; they in Europe have organized class. If caste has produced certain evils, class has not been productive of anything less.

> If class helps conserve certain social virtues, caste does the same in equal, if not greater degree. (Gandhi 1965: 6)

On the relations between industrialists and workers he wrote: 'The relation between mill-agents and mill hands ought to be one of father and children, or as between blood-brothers' (Gandhi 1966: 207). He hoped he and his followers would be able to change the capitalists' and landlords' hearts. He did not want to dispossess them. In his future social order, they would function as benevolent trustees (Gandhi 1966: 236–9).

This is not the place to make a thorough analysis and evaluation of Gandhism's relevance for today.[10] But it should be clear that eco-socialism, the scientific socialism for the 21st century, cannot be based on Gandhism.

ECO-SOCIALIST POLITICS FOR TODAY AND TOMORROW

Analysis, theory, speculation – all this is necessary. But from it we only have a vision, which is of little use unless there is also action to realise it. What can and should be done today and tomorrow, so that we at least come closer to beginning the transition?

I believe I have achieved the *theoretical* synthesis; now the *practical* synthesis of concrete ecological and socialist politics, of the two movements, must come. A new eco-socialist movement must be built. This is primarily the task of activists in these two fields. But I would like to make some fundamental remarks on this question – as a contribution to the practical synthesis.

All eco-socialist activists must realise one fundamental difference between our movement and the social and socialist movements of the past. Until the ecology movement emerged, most large movements arose from social problems. In earlier epochs, in the industrial societies, most social problems could be (at least partly) solved through economic growth. Acute poverty could be overcome and wages increased. The poor, the unemployed, and students got financial help. Care of the old and the ill was provided for. Women got the franchise and jobs in industry and trade. Democratic rights were recognised and extended. The demands of all social movements could be fulfilled to a large extent, thanks to the growing cake. But with the emergence of the ecology movement, the situation has changed completely. Not only must the cake not grow, it must shrink. The very basis of the ability of industrial societies to solve social problems must be attacked if the problems from which the ecology movement arose are to be solved. For the first time in history, a social movement 'promises' a lower standard of living.

For radical socialists, especially for communists, with their tradition of thinking in terms of revolution, a quotation from Walter Benjamin may serve as an eye-opener:

> Marx says revolutions are the locomotive of world history. But perhaps it is entirely different. Revolutions are perhaps the attempt of humanity travelling in a train to pull the emergency brake. (quoted in Fetscher 1980: 8)

They can now join the eco-socialist movement with the conviction that a revolutionary task is waiting for them. I can also show that capitalists are afraid of this synthesis. In 1990, in an article on the protection of the global atmosphere, the German newsmagazine *Der Spiegel* reported: 'The assistants of President Bush suspect that protagonists of a planned economy, defeated all over the world, are now striving to achieve via environmental protection what they failed to achieve during the cold war: the victory over capitalism. Thus Darman ... incited fear of "radical greens" ... ' (*Der Spiegel*, 21 May 1990: 163). Indeed, a serious ecology movement is today the greatest enemy of capitalism.

But radical socialists and communists would have a big problem after they have learnt the real ecology lesson and become eco-socialists. They would no longer be able to fight for the material interests of the working class in the same simple way as before. They would have to tell workers – undoubtedly in the first world, and perhaps also in the organised sector in the third world – that their standard of living is too high for the health of the environment, for the survival of other species, and for the rights of future generations.

They would have another big problem. They believe(d) that with the proletariat, for the first time in human history, a class came into being whose interests were identical to the interests of all humanity, and that the proletariat worldwide could and would unite. But since there are limits to material growth, these two beliefs are not correct. We know that there is no congruence between the material interests of the proletariat in the first world and in the third world. Why should a German worker demand that plantation workers in Latin America get the same wage as he gets? If this demand were fulfilled, banana prices in Germany would skyrocket. Workers in the first world have much more than their chains to lose, they have their prosperity.

Theoretically, of course, we can differentiate between material interests and some hypothetical 'real' or 'deeper' interests, as well as between short-term and long-term material interests. But 'deeper' interests are vague, and in the long run we are all dead. Eco-socialists in the first world cannot even point to the health hazards of some industries and of toxic wastes, for such things can be transferred to and dumped in the third world, easily and at low cost.

We must therefore conclude that if we base our efforts to win people for our movement on their short-term material interests, we will necessarily counteract our goals. Only in single-issue environmental campaigns, such as against a particular polluting chemical, does the aim coincide with people's material interests. Even then, the campaign would oppose the material interests of the workers in the factory concerned.

This means that if the movement forms a party and takes part in elections, it must, for many years to come, be prepared to be rejected by the vast majority of voters. The temptation would be great to water down both analysis and programme in order to get a few seats in the parliaments or a few ministries. But 'successes' achieved in such a way would transform the party, as they transformed the German Green Party (see Sarkar 1994). Eco-socialists must not tell voters that they can eat the cake and have it too.

Let me point out a mistake made in this matter by Ryle. Referring to the 'violently disruptive effects' of the policies of the Conservative government in the UK 'on local and sectoral economies: steel, coal-mining, ship-building, merchant shipping', he asks the rhetorical question: 'If the Right has been able to win political support for these in the name of a rather abstract "realism", can an ecologically informed socialism not aspire to creating a consensus for its own very different objectives?' (Ryle 1988: 66). Of course, I share his hope. But the parallel is wrong. The Right promised continuation of material prosperity through their kind of realism – not, of course, for everybody, but on the whole for the British people. And they succeeded in attaining their end. Eco-socialists would not make such a promise. On the contrary. That is why their task would be much more difficult than that of the Tories. It will be the most difficult of all tasks in the next century.

A few lines further on, Ryle mentions, as an example of 'ecologically informed socialism', the well-known Lucas and Vickers workers' plans for alternative production. But eco-socialism will not be achieved by mere product conversion, as it makes little difference to the ecology or the rate of resource depletion, although in other respects there is indeed a difference between producing tanks and producing trucks.

I think that at present the emphasis should be on an eco-socialist movement and not on a party in its usual sense. If a party is formed, its task should be to use elections to strengthen the movement. The moral growth of society at large, *the* precondition of success, must be the primary purpose of the movement at present. Readiness to sacrifice luxury and comfort must be generated. For that, a simple-lifestyle campaign is essential. It must be based on ecological arguments, appeals to people's sense of equality, justice, and solidarity with the third world, and the interests of the future generations.

The movement must also fight for job security and social security. Job sharing and reduced working hours must be campaigned for as short- and medium-term measures against unemployment. It must, however, be said openly and clearly that the long-term solution lies in labour-intensive technologies. This aspect of the movement is very important, for unemployed and otherwise insecure people generally find it difficult to practise solidarity or to give priority to protecting the environment.

The movement must mount a continuous campaign for some retreat from the world market and for a certain degree of protectionism – on the basis of both ecological and employment arguments. Also in respect of the purely home market, the movement must demand that the state control capital and the market more strictly. An example is to demand stricter regulation of rented housing.

There is no need to go into more detail. Experienced activists in the two movements would know what had to be done once they had accepted the basic eco-socialist positions. They would have no illusion that their movement and campaigns would suffice to change society, so to speak, from below. But they would hope that in a few years the movement and the campaigns would have prepared the ground for the formation of an eco-socialist government. Even if that does not come soon, other parties and politicians in power might take over some eco-socialist positions, at least in part. The readiness of the majority to accept sacrifices would also be seen by such parties and politicians, who would feel under pressure.

Two more points need to be made. It is unlikely that the eco-socialist movement could succeed in any one small country. In 1917, it was difficult for the Bolsheviks to defend their revolution in the huge Russian empire. Today and tomorrow it would be much more difficult for the eco-socialists of a small country to defend their regime, even if they came to power with the support of a majority of citizens. No military intervention by hostile neighbours would be necessary to put an end to the regime. In today's vastly increased global economic interdependence and interlacement, compared to 1917, a small country with a newly elected eco-socialist government could easily be plunged into severe economic difficulties, through boycotts, trade embargoes, and so on, so that a planned, orderly retreat could not even begin. In large countries, such as India, Brazil, the USA or Australia, it might be less difficult. But the best strategy would be to build an eco-socialist movement first in large regions, such as western Europe, and not try to come to power before the pressure for change in the direction of eco-socialism has built up in several countries. To say more than this would be too speculative at this point.

Socialists and communists who have learnt their ecology lesson must finally drop all illusions as to the revolutionary or leading role of the proletariat. There is absolutely no special reason why the proletariat should support the changes eco-socialists want to usher in any more than white-collar employees (clerks, teachers, engineers, sales staff, and so on). The former are as much addicted to consumerism as the latter, and the latter are today suffering as much from unemployment and ecological degradation as the former.

Erich Fromm came to the conclusion that today there are 'only two camps: those who care and those who don't care' (Fromm 1979: 196). Those who care can be found in all classes or strata of society – except perhaps among entrepreneurs, who must demand economic growth, without which they are in danger of going bankrupt. Eco-socialists would be the most active among those who care.

NOTES

1. Social consumption is what the state spends on social, cultural, educational, medical, and other similar infrastructures that the citizens use without having to pay for them directly.
2. Pepper obviously means hydrogen obtained from water.
3. For a thorough critique of Gorz's views, see Sarkar 1990.
4. Ryle surely means limits imposed by ecological responsibility.
5. For those who do not read German, I have published a long review essay on Ullrich's book and its subject matter. See Sarkar 1983.
6. Federal Republic of Germany (West Germany) and the German Democratic Republic (East Germany).
7. My views on the population problem have been elaborated in Sarkar 1993b.
8. The Polish anarchist Waclaw Machajski warned against this probability before the October Revolution in Russia (1917) (see Kolakowski, 1978: 162).
9. Marx's own formulation is: 'free association of solidary individuals'.
10. For such an evaluation see my article 'Von Gandhi lernen – aber wie?' (Sarkar 1988).

CHAPTER 7

What about Progress?

In this last chapter, I want to deal briefly with a few questions arising from my conception of eco-socialism and some associated questions not yet dealt with.

DEVELOPMENT OR PROGRESS?

It is clear that, as far as the economy is concerned, eco-socialism means almost the opposite of what is usually called development. In order to avoid this logical conclusion, some radical ecologists have redefined 'development'. Daly, for instance, writes:

> Growth means a quantitative increase in the scale of physical dimensions of the economy. 'Development' means the qualitative improvement in the structure, design and composition of the physical stocks of wealth that results from greater knowledge, both of technique and purpose. A growing economy is getting bigger, a developing economy is getting better. An economy can therefore develop without growing, or grow without developing. (quoted in Sivaraksa 1996: 70)

This is all right. But unnecessarily changing the meaning of a concept creates only confusion. For Daly's kind of development, one can use instead 'improvement' or 'betterment', terms which are defined in the above quote.

But no thinking person is concerned only with the economy. The organisers of the congress on 'Ecology Between Self-Limitation and Emancipation' feared that ecological politics could only mean 'de-development of society' (see pp. 204–5). It would have been clearer if they had spoken of retrogression of society. Anyway, since for Marxists, other leftists, and most other people, human and social progress is absolutely dependent on the development of the productive forces – that is, on economic development – they may logically conclude that any kind of low-level steady-state economy would necessarily mean the end of all progress. But is this conclusion right?

In the preceding chapter, I have argued with Ullrich that socialism has no necessary connection with the level of development of productive forces because it is mainly a question of human relations. Since socialism is possible on the basis of a low-level steady-state economy, I assert also that progress is possible on this basis. We have only to clarify what progress means.

As the first step towards this clarification, I shall first quote two Western political economists. Keynes wrote in 1931 that he hoped that '... the Economic Problem will take the back seat where it belongs, and that the arena of the heart and head will be occupied, or re-occupied, by our *real problems* – the problems of life and of *human relations*, of creation and *behaviour* and religion' (Keynes 1931: vii; emphasis added). Keynes, of course, thought that this would happen because the Western world had the 'resources and the technique ... capable of reducing the Economic Problem' (ibid). None the less, it is worth noting what he considered to be the real problems. Economic growth was not one of them. It is also worth noting that he considered the resources and the technique that the Western world already had in 1931 to be sufficient; the only thing still needed was 'the *organisation* to use them' (ibid; emphasis added).

The other political economist, John Stuart Mill, advocated a 'stationary-state' economy in the 19th century.

> It is scarcely necessary to remark that a stationary condition of capital and population implies no stationary state of human improvement. There would be as much scope as ever for all kinds of mental culture, and moral and social progress; as much room for improving the Art of Living and much more likelihood of its being improved ... (quoted in Daly 1980: 15)

Mill was not alone in this view in the 19th century. The French socialists – above all Fourier, Proudhon and Blanc – were 'almost reactionary in their progressivism, for they tended to see the possibilities for improving society in the reversal of some of the modern industrial trends' (Sklair 1970: 49). William Morris's vision was called 'regressive' utopianism (Ryle 1988: 23).

What Keynes and Mill say in the above quotes is very general. We can make the elements of progress that would be possible in a model eco-socialist world a little more concrete, without discussing all of them in detail. Above all, the possibility of peace, both within a state and between states, would increase. In an eco-socialist society, since every able-bodied adult would be meaningfully employed, there would be much less crime and violence than today. Since equality would be one of the highest principles, there would be no class conflict; exploitation and oppression would not exist, nor the subordination of women to men. Since demand for scarce resources would diminish, the threat of war between states on account of them would also diminish. Reduced competition in the world market would make friendly relations between the peoples of the world possible. Stable populations and sustainable economies would end large-scale migration from today's poor to today's rich countries, thus reducing xenophobia in the latter.

Complete literacy in all countries would be possible without building large schools, as in Europe today. There would not, of course, be a personal library in every household, but more readers would use public libraries. More educated people would take more part in public discussions. There is no reason to think that there would be less music, art, or literature in an eco-socialist society. But it is probable that there would be fewer full-time professional musicians, artists, writers, composers, and so on. Most such people would probably need a second, part-time job.

It would be possible, as shown in the previous chapter, to extend democracy and participation to the economic sphere. There is no doubt that freedom, to the extent that it is understood as a result of high material consumption, would be more limited than in today's advanced industrial societies. Moreover, duties towards society would have to be fulfilled. Otherwise, there is no reason why civil liberties, such as freedom of speech, should be any less extensive than today.

On the whole, there would be great progress in human relations and behaviour, and in relations between peoples and states. At least, the possibility of such progress would increase. In India, for example, an eco-socialist regime would be more likely than any other kind of regime to succeed in abolishing the caste system and the institution of dowry.[1] That would be enormous progress for Indian society.

ALL HUMAN RIGHTS FOR ALL

An important element of progress should be the realisation of human rights. In 1776, it was declared in section 1 of the 'Virginia Bill of Rights' 'that all men are by nature equally free and independent and have certain inherent rights ...'.[2] In 1948, the UN recognised in the preamble of its 'Universal Declaration of Human Rights' 'the inherent dignity' and 'the equal and inalienable rights of all members of the human family' to be 'the foundation of freedom, justice and peace in the world'. But we know that not all members of the human family enjoy these rights.

Most human rights organisations, such as Amnesty International, protest only against *violations* of *some* human rights. Strangely, they do not protest against the *non-realisation* of some other centrally important human rights; the declarations themselves contain hardly any thoughts as to how these rights might be realised.

In article 1 of the UN declaration one reads: 'All human beings are born free and equal in dignity and rights'. In article 1 (1) of the 'Basic Law' of Germany one reads: 'The dignity of humans is inviolable'. But we know that all humans are not born free and equal. Some are born rich, some poor, some free, some unfree. It is also not true that the dignity of humans is inviolable. It is violable. A society in which such statements correspond to reality has not yet existed. The authors of such statements should, therefore, have written *how* their goals could be attained. They have defined equality as merely equality before the law. But does the law anywhere at least promote real equality between citizens?

The UN declaration says: 'Everyone is entitled to a social and international order in which the rights and freedoms set forth in this Declaration can be fully realized' (article 28). Few would deny that the economic order is the most important component of any social and – for the last few decades – international order. But neither the various declarations of human or fundamental rights nor the laws of any existing state say that all humans should be *economically* equal.

Economic equality is a value in itself and ought to be the most important component of the principle of equality. Moreover, without this equality it is very difficult to realise other equalities and freedoms. A person without financial strength cannot, for example, stand for election to the post of President of the USA. That is, he or she may formally be allowed to, but would not have an equal chance to bring his or her programme and personality to the notice of the electorate. He or she could not go to court to get an injustice redressed, nor publish a local newspaper. Someone who has to work eight hours a day in a factory does not even have the time and energy to write a letter to the editor. And a worker, who has nothing to sell other than labour power, must often, especially in times of high unemployment, tolerate the violation of his or her dignity by the managers or owners of a company if he or she does not want to risk being fired.

Economic equality does not even exist as a goal in the constitutions of so-called democratic states. On the contrary, almost all declarations of human or fundamental rights ceremoniously proclaim the right to property. The French constitution of 1791, adopted just two years after the French Revolution with its slogan 'liberty, equality, fraternity', even proclaimed this right to be 'sanctified' (article 17).

It is obvious that, in a world with finite resources, not everybody can own property in the sense meant by these declarations and constitutions. Theoretically, it would be possible to give all people an equal legal title to the resources of the world or a state. But then the 'democratic' states, about whose constitutions we are talking, would not have come into being. If a big landlord insists on his or her right to property, then most fellow inhabitants of the village cannot become owners of agricultural land. If the mineral and forest resources of a region are the property of a few, then other people in the region cannot own them.

Actually, the most important condition for realising other human rights could be fulfilled if everybody had *sufficient* economic strength. The UN declaration proclaims some 'indispensable' economic rights (article 22) of every member of society. It proclaims the 'right to work' (article 23/1), the 'right to just and favourable remuneration' (article 23/3), the 'right to a standard of living adequate for the health and well-being of himself and of his family, including food, clothing, housing and medical care' (article 25/1). But not even all citizens of the rich democratic states of the North enjoy such rights. Several million are homeless in these states, and thousands have to sleep in parks and on footpaths. Every winter, several dozen freeze to death. Equality between men and women has not yet been achieved. Racism has become rampant.

It cannot be argued that everything needs time. If more than 200 years of development have not been sufficient to realise human rights, there must be some fundamental defect in the concept. It is the non-existence of the right to economic equality. Protagonists of capitalism say openly that this right cannot be realised. They say that egalitarianism is bad for economic growth, for it allegedly kills initiative, entrepreneurship, and the will to perform. Even social security systems, where they exist, are being gradually dismantled. We must then conclude that there is a contradiction between human rights and capitalism.

This contradiction has always been there, and the authors of the various declarations of human rights were surely aware of them. They could not declare economic equality to be a human right, for the idea of human rights, with its strong emphasis on the right to property, arose through an historical process, in which large sections of humanity were robbed of their properties and resources by the very European and American bourgeoisies who made the bourgeois-democratic revolutions of the 18th and 19th centuries. This robbery took place through conquest of colonies and, at home, through processes such as the enclosure of commons. Neither the authors of the Virginia Bill of Rights nor their descendants could have acquired any property without first killing or driving away the people who lived in America before their arrival. In the process, large parts of humanity were also deprived of their personal freedom – the slave trade and *apartheid* were only two of the worst cases.

In the late 19th and early 20th centuries, the ordinary people of the colonial powers also developed an interest in colonies. In Germany, the majority in the Social-Democratic Party, the party of workers, supported the state's policy of colonial expansion; some wanted a peaceful partitioning of the world among the European 'culture states', others had no objection to the use of military means (cf. Mandelbaum 1974). In the post-colonial era, exploitation of the former colonies continues through other means, neo-colonialism, which Johan Galtung calls 'structural violence' (Galtung 1975).

The human rights organisations report violations of human rights in the third world. But they do not say a word about the true causes of these violations. The ruling élites there do not torture, murder and incarcerate dissidents because they are sadists. They do it because they would not otherwise be able to maintain their political power to continue their exploitation of the majority of the people. And it is not only dictators: democratically elected governments – for example the government of Turkey, the West's NATO ally – also do it when necessary.

'Democratic' states of the West support and protect dictators and human rights violators such as Mobutu in Zaire and Pinochet in Chile because the latter serve their interests. They also drop dictators when they cease to be useful. Their chief interest is to maintain the world order in which they dominate, which guarantees that they continue to enjoy the lion's share of world resources, including the labour power of the people of the third world. They protect dictators and other authoritarian rulers because only such rulers can ensure that IMF Structural Adjustment Policies are implemented, that interest on credits from the West continue to be paid, and that Western investments in the countries concerned are secure. It is also a fact that ordinary people in the Western industrial countries have an objective interest – whatever their subjective feelings may be – in the continuation of this world order, because Westerners, even the poor and unemployed, benefit from it.

Protest against violations of human rights in the third world remains empty unless the root causes are attacked: these are the present social, economic and political order in every country, and the whole world order. There are strong reasons to assert that in an eco-socialist regime, and even more so in an eco-socialist world order, the chances of realising human rights would be much greater.[3]

ONE WORLD, OR DIFFERENT CULTURES AND CIVILISATIONS?

The human rights declared by the UN in 1948 are supposed to have universal validity. This presupposes the concept of one world and one humanity. The declaration also uses the term 'the human family'. This concept is shared by all who have overcome nationalism and are working for peace, development, and progress in the whole world, and of course by all socialists. But in the last few years, it has been rejected by many ecologists and proponents of a different kind of development.

Progress, or back to traditional cultures?

Some critics of development (in fact, only the hitherto dominant models) have introduced cultural arguments. I share their rejection of development, but not what they say regarding traditional or indigenous cultures. I am afraid that they are unwittingly supporting movements harmful to humanity.

The cultural approach

I think we must today talk of two approaches to the critique of development: the cultural and the ecological–economic–political. Protagonists of the first do not wholly ignore the second. But it appears that for most of them, the principal evil in development is that it destroys or suppresses the traditional cultures of the South.

Wolfgang Sachs (1989) maintains that the West[4] imposed or palmed off on the South the idea and programmes of development, through which the whole world was to be made one unit and the ways of life of the various peoples uniform. This development has, according to him, caused the 'evaporation' of cultures and languages and destroyed the contented subsistence economies of traditional cultures. The result, according to him, is that these peoples have not only become poor but have also lost their cultural identity.

For Sachs, the slogan 'one world' is a horror, because it endangers the 'self-willed ways of living and understanding' and because it restricts the space for self-determination and autonomy of the peoples of the world. He is aware of the dangers to the biosphere and of the logic of 'spaceship Earth'. But his love of 'difference' and 'chequeredness' is very strong, and hence it is for him 'a sacrilege to design the global space as a united, highly integrated world'.

T.G. Verhelst (1990) speaks of 'the right of peoples to be different'.

Whereas Sachs is against even literacy programmes for tribal peoples who do not have a written language, Verhelst does not oppose development. But he criticises programmes for grafting the Western model of development, including Western technology and culture, on the peoples of the South. According to him, this is why many local partner groups reject or remain passive towards development projects promoted by organisations from the North. He interprets this as a form of resistance to imperialism. He thinks that the 'failure to give due recognition to the indigenous cultures is one of the fundamental reasons for the failures and difficulties of development work'. He demands respect for all cultures and thinks that those NGOs that understand development work as a contribution to liberation should consider indigenous cultures as the foundation of development.

Such views have also been expressed in the South. Let me give a few examples from India. Smitu Kothari writes: 'And alongside this biological survival of marginal communities is the continuous threat to the survival of their cultures. The undermining of cultural plurality is thus built into the dominant model of 'progress' (Kothari 1985: 8). For Anil Agarwal and some ecologists of the South, an examination of the cultural and biological diversity of human society resulted in 'an understanding of the *essential rationality of each culture*, how it had emerged in its own particular ecosystem and the way people had developed social and production systems, behaviour patterns and survival techniques. Once this was understood, there was an immediate respect for *all* cultures ...' (Agarwal *et al.* 1987: 351; emphasis added). Some Indian ecologists maintain that it was due to its religious and cultural heritage that India's ancient (or pre-colonial) economy was very ecological and that *ahimsa* (non-violence) and *dharma* (right conduct) are parts of that heritage.

S. Kappen (1994) even demands that 'the right to ... culture identity must be affirmed as a fundamental human right'. He asserts that 'a development that consists in the satisfaction of our ... needs will necessarily be culture-specific'. In his opinion, 'culture must body forth not only into the political organisation of society', but also into 'science'.

Cultural identity is a trap

For reasons of space, the foregoing could only be a short and approximate summary of the views I want to criticise here. It illustrates a trend. It is not always clear, firstly, whether the author refers only to traditional cultures of tribal and marginalised peoples or also to those of the majorities among the peoples of the South. Secondly, it is not always clear whether one means by traditional culture the views and ways of life expressed and recommended in the ancient holy and philosophical texts or (also) the cultures actually practised in the past or the present. I have the impression that everything can be meant that is not Western or Westernised.

It must be emphasised that participants in such discussions use the word 'culture' in its social-anthropological meaning, namely 'that complex whole which includes knowledge, belief, art, morals, law, customs, and any other capabilities and habits acquired by man as a member of society'. In this meaning, 'culture' includes also 'the material organization of life', that is, 'social and economic institutions' (Edwards 1967: 274).

My chief criticism against the cultural approach is that it uses arguments that are not only wrong but also harmful. Development should be rejected, but not because it is Western or foreign; it should be rejected because it is ecologically, economically, politically, and socially untenable and harmful, as much in the West as anywhere else. Most protagonists of the cultural approach would be likely to agree to the second part of that statement, and therefore their cultural approach is unnecessary for their rejection of development.

But their approach is also harmful, and not only because it romanticises traditional cultures, producing a false image of reality. It becomes clear where it can lead when we read the following in a report of a conference on the new world order held in Germany (in free translation):

> Hans May, director of the Protestant Academy, agreed that even in the next few decades it would be impossible to equalise the standards of living in the Western and Eastern (former GDR) parts of Germany, and accepted the idea that this impossibility should be compensated for by strengthening the regional identities of the Eastern parts. He then proposed this as a model for the whole world. He said that the impossibility of achieving a standard of living in the South which is approximately equal to that in the North leads us to conclude that *human rights must be differentiated and regionalised*. In this connection, May cautioned *against discrediting religious fundamentalism*. (*Frankfurter Rundschau*, 27 April 1992; emphasis added)

Here the motive is not respect for all cultures, nor love of difference, but fear of the South's aspiration to catch up with the North economically, which would be ecologically disastrous for the whole world. Moreover, at the UN Human Rights Conference (Vienna, June 1993), all groups from the South insisted on the universality of human rights. Their slogan was 'all human rights for all' (cf. Sethi & Sheth 1993). It is only the rulers of some countries who want to reject this universality with the argument of cultural difference. Obviously, they too are afraid of the aspirations of their own exploited and oppressed and women to catch up with the rich, the middle classes and the men. I am not suggesting that all protagonists of the cultural approach explicitly share Hans May's conclusion, but that is the direction in which their emphasis on cultural identity and difference logically leads.

It is not correct to say that the North imposed or palmed off its development models and culture on the South. Most people in the South gladly accepted at least the greater parts of development and Western culture. In India, a National Planning Committee had been constituted by 1938 at the instance of the Indian National Congress. The objectives of planning were '... to raise the general standard of living of the people as a whole ... by the development of the resources of the country to the maximum extent possible' (quoted in Majumdar *et al.* 1967: 969). And since the early 1980s, some European politicians have been explicitly advising the South not to copy the West – perhaps out of the same fear that Hans May has. Edgar Pisani, a former leading politician of the European Economic Community (EEC) said: 'Divesting our relations [with the South] of any hint of racism is to affirm: we are different and we are going to stay different' (quoted in Verhelst 1990: vii). Narducci, then President of the EEC Parliament, spoke of the 'pernicious conviction that there exist superior cultures that must be asserted' and demanded 'respect for specificity and cultural identity' (quoted in ibid: 149–50). In the Third Lomé Convention (1984), the EEC as a whole promised to 'promote the cultural identities' of the peoples of Africa and the Caribbean and Pacific regions. It opined that development should be 'based on their cultural and social values' (quoted in ibid: 150).

We have seen what brutalities and oppressions religious fundamentalism can lead to. One could argue that a cultural approach does not mean a religious approach. But religion is a very important element of culture, and cultural identity is, to a large extent, religious identity. It does not help at all to obfuscate the matter by introducing unsound differentiations, as Kappen does: 'While ... religious revivalism and fundamentalism is cropping up in many parts of the world – which itself is a pointer to a loss of wisdom – people are losing their moorings in authentic religion' (Kappen 1994). What is authentic religion? Are not the *Sharia*, the Islamic law book, and the papal decrees part of authentic Islam and authentic Catholicism? Is not the caste system a part of authentic Hinduism? All three religions have their infidels: *Kafirs*, heathens, *Mlechchhas*.

But it is immaterial what one means exactly by 'cultural identity': laying emphasis on it causes separation between peoples, nations, ethnic groups, and so on. Of course, under certain circumstances it can unite a group of people and so generate internal solidarity. But it can do so only by drawing distinctions between this group and the others. Under unfavourable circumstances, the search for identity can easily end in a search for enemies. In Northern Ireland it is a mixture of religious and national identities; in the former Yugoslavia it is a mixture of ethnic and religious identities; in the former Soviet Union it is a mixture of national, racial and religious identities; in the USA it is racial or colour identity; and in India, in addition to all that, it is often also a linguistic one – Bengali, Assamese, Oriya, Tamil, Hindi, and so on. Of course, in such conflicts there is almost always a deeper economic cause, but emphasising separate cultural identities makes finding a solution more difficult.

For India, emphasising cultural identity constitutes a great danger. There is no such thing as a traditional Indian culture. There is a traditional Hindu culture, but that is not the traditional culture of Indian Muslims, a very large minority, or of Indian Sikhs. If Kappen's idea becomes reality, it would be traditional Hindu culture that would 'body forth ... into the political organisation' of Indian society. That would be the end of the secular Indian state, and we would have a Hindu state, with consequences that need no elaboration.

The opportunistic attitude of some in the West, such as Hans May, towards religious fundamentalists and cultural and other kinds of identity fetishists, will be of no use. For in economic matters, the latter generally do not value any difference: they all want to catch up with the North and contribute as much to the degradation of the biosphere as the industrial societies of Europe and America. It therefore surprises nobody to see Iranian *ayatollahs* processing holy scriptures with computers in the library at Qom. For the North, industrial wealth; for the South, cultural identity – this new ideology of imperialism is easy to see through.

'We are different and we are going to stay different' – this ideology must be rejected for another reason. With it, Americans can defend the ecologically destructive and exploitative 'American way of life' as their cultural identity; Hindus can defend untouchability and the caste system; and almost all peoples of the world the subordinate status of women. These are indeed parts of traditional cultures. In South Africa, the focus on different cultural and ethnic identities enabled even many liberal and nationalist–Christian whites to support *apartheid* policies without having to use the discredited term 'race'.

There are also people who reject development for ecological, economic and social reasons, but nevertheless land in the culture trap. Here is an example: in the Singrauli region of Madhya Pradesh, India, some activists defending the interests of victims of development expressed the view that the only alternative to development is a return to the past. They spoke of the glory that was India before the British came. They maintained that the villages of ancient India, and of precolonial India, were a happy world, free from exploitation, that there were no famines then, that women used to be treated as equals of men. Although these assertions are all wrong, they express some ideals. But the culture of ancient India also contained caste discrimination, so these romanticists also defended the caste system (cf. Dhagamvar 1985). Another example is Kappen, who convincingly criticises capitalism and industrial civilisation from ecological, political and social standpoints, but his solution is also 'a retrieval of the perennially valid insights of the past', whatever that might mean.

The alleged wisdom and rationality of traditional cultures

One meaning of 'perennially valid insights of the past' might be the alleged ecological wisdom of the ancients. Without doubt, most cultures before the industrial revolution had found some sort of ecological balance. Otherwise they could not have survived. They must also have had some ecological wisdom, for, after all, their livelihood depended to a much larger extent than today on the health of their ecology. But they were also blessed by certain circumstances: firstly, they did not have the technologies of today, nor the fossil fuels, with the help of which modern cultures are rapidly degrading the biosphere. Secondly, the pressure of population growth was much less. Thirdly, there was enough thinly populated territory to which that part of the population that could no longer be fed could migrate. Europe would have been ecologically ruined if America had not been discovered in the 15th century.

But archaeologists and historians have unearthed much evidence to show that the ancients also destroyed their ecology in many parts of the world. But the process was very much slower than today, so that not many individuals could perceive it in a lifetime. Clive Ponting (1991: chapter 5) gives several examples: the Mediterranean region; the Maya, the Sumerian, and the Indus Valley civilisations; the catchment area of the Yellow river in China; Greece in the days of Plato. The causes were deforestation, overgrazing, overpopulation, over-irrigation leading to waterlogging and soil salination, agriculture on unsuitable land leading to soil erosion, and so on.

It cannot even be said that all human groups that lived before the rise of any civilisation had an eco-congenial way of life. There is evidence to show that in Australia and America, the earliest settlers caused massive disruption to the eco-system. Over-hunting, especially the technique of fire-drive hunting, caused the extinction of 73–86 per cent of large mammals (ibid: 35; Lovelock 1987: 133).

In present-day ecological literature, we often read about the ecological wisdom of pre-industrial peoples. One popular, romanticised example is that of the Bishnois, some 300 of whom sacrificed their lives in order to protect trees (cf. Sunderlal Bahuguna's account in Tüting 1983: 16–19). Actually, these Bishnois had a material interest in protecting trees that protected their village from the Thar desert of Rajasthan, India. But the fact cannot be overlooked that the king and the axemen who wanted to fell the trees also had a material interest in doing so, namely to get fuel to produce lime, which they needed to build a palace. It was simply a case of conflict over resources in pre-industrial, pre-colonial India. In this respect at least, there is little difference between the past and the present, except in tempo. Ecologically, the past was no wiser than the present. There is no proof for the general validity of 'grandfather's law'.

Cases of ecological destruction revealed by archaeologists and historians are perhaps not very numerous. But there is no doubt that in all traditional advanced cultures of the world – the Indian Hindu, the Arab Islamic, the European Christian, the Chinese Confucian, and so on – there has always been exploitation and oppression, violence and crime. We know that in almost all cultures of the world, hierarchy, a class or caste system, and patriarchy have condemned large parts of the population to a permanently humiliated existence for no other reason than that they were born in the wrong family or the wrong gender. And almost all peoples of the world, including the tribal (with the possible exception of the most primitive hunters and gatherers), have waged wars. It is not even true that the wise men of ancient Indian culture were apostles of non-violence. The entire *Bhagavat-Geeta*, the most important Hindu religious text, is an exhortation to fight a war. The wise men and religious texts of the Jews told them that they were the chosen people of God, and so could conquer the lands of others. Buddhism is a non-violent religion, but it does not forbid the economic exploitation of fellow human beings. Sulak Sivaraksa, the Thai social activist, writes:

> Institutionalized Buddhism (with a big B) explains oppression with *karma* [as does Hinduism]. It says that both peasants and landlords reap the fruits of the deeds of their previous lives – the peasants the fruits of their bad

> deeds and the landlords the fruits of their merits acquired through the construction of temples and statues of Buddha. Both the rich and the poor are encouraged to support institutionalized religion and the amassing of material wealth in the brotherhood of monks for their own well-being in their next life. (Sivaraksa 1996: 65)

Institutionalised Christianity – notwithstanding the Sermon on the Mount and the concept of a loving God – has led crusades, supported wars, amassed material wealth, and has not objected to exploitation and oppression.

In view of these facts, one cannot have respect for *all* cultures. Except in the case of those economies that were probably more or less ecologically sound, I see no 'essential rationality' in traditional or pre-industrial cultures. The ancient Hindus did not create their caste system and untouchability in order to adapt themselves to their particular ecology. And nowhere in the world does the ecosystem make it necessary for men to oppress women.

In almost all societies there have been contradictions and conflicts, and some of them have been related to cultural values. There have been struggles against exploitative, oppressive, discriminating values, mores and customs, struggles against superstition and for scientific knowledge. And, in general, there have been struggles for emancipation. No culture deserves wholesale respect or wholesale contempt. But all of them are inadequate for today's difficult, complex tasks. Today, only dialogue and critical solidarity are appropriate behaviour in relations between peoples of different cultures.

To be sure, there have been tribes in which solidarity and mutual help has been the norm. But this has never been extended to neighbouring tribes; there has been no norm to prohibit their exploitation or even massacre. North American Indian cultures of previous centuries – which enjoy among many leftists and eco-activists the reputation of being ecological and social models – were not particularly peace-loving. Citing the historian Dickason (who is partly of Amerindian origin), David Orton writes:

> warfare and hostilities between tribes was endemic throughout the Americas. The Iroquoians practised cannibalism and torture. Some tribes made human sacrifices. On the Northwest Coast [of Canada], class divisions were based on wealth, with hereditary chiefs, nobles, commoners, and slaves who could be killed, etc. (Orton 1995)

Similar traits were evident in the cultures of the Aztecs and the Incas (cf. Rügemer 1992). In our own day, the conflicts and massacres among Hutus, Tutsis, Zulus and other African tribes are well known.

In any case, it is illogical to build up an ideology or alternative paradigm on the basis of a few positive elements in a few exceptional cultures. It is equally illogical to quote a few suitable sentences from rarely read holy and philosophical texts of traditional advanced cultures and then to assert: thus was their culture. The *Bhagavat-Geeta* and the *Upanishads* were no more expressions of actually existing Hindu culture than was the Sermon on the Mount an expression of actually existing Christian culture.

Theoretical error

A fundamental theoretical error of the protagonists of the cultural approach is to draw a parallel between ecology and culture. They talk of ecological and cultural diversity in the same breath. Diversity is absolutely essential and monoculture very bad for the health of the ecology of a given area. But for the health of a human society, it is not bad if people speak the same language or wear similar clothes or worship the same God. There is another difference. A mango tree will not grow in Europe, nor an apple tree in south India. But Christianity has thrived in the Philippines and *yoga* has become very popular in Europe. We must oppose genetic engineering because of the danger of irreversible damage, but an artificially created cultural or social institution can be abolished or counteracted if it proves harmful.

Lewis Mumford, one of the major philosophers of the eco-alternative movements in the West, gives another reason for preserving what he calls 'national and regional' cultures:

> One of the great benefits of individualized national and regional cultures is that, if the opportunities are consciously seized, these potential alternatives can be experimented with under varied conditions and their advantages compared. Any philosophy of history that takes account of natural and human diversity must recognize that selective processes in nature have reached a higher stage in man, and that any mode of organizing human activities, mechanically or institutionally, which limits the possibilities of continued trial, selection, emergence, and transcendence, in favour of a closed and completely unified system, is nothing less than an effort to arrest human cultural evolution. (Mumford 1970: 159)

This is not convincing. Firstly, Mumford is reducing national and regional cultures to the status of guinea-pigs, to be experimented with. Secondly, the argument here sounds as if we should have a culture bank just as we should have a gene bank. Neither of these is a very respectful attitude to other cultures. Thirdly, it is the protagonists of the cultural approach who are trying to arrest human cultural evolution by demanding that traditional cultures should remain as they are or even become again as they were before. My argument against Mumford's observation is that neither cultural guinea-pigs nor a culture bank is necessary for his purpose. In nature, if a species becomes extinct, it is gone for ever. But we can, if we wish, revive old or extinct cultures, or elements thereof. Descriptions of old cultures and their various elements are available in books. The revival of the Olympic games is a case in point. Moreover, whereas genetic engineers need a gene bank for their experiments and creations, we can use our fantasy and intelligence to create entirely new social or cultural institutions, rules and regulations.

What do the peoples want?

Protagonists of the cultural approach insist on peoples' right to be different, the right not to become like Europeans. But who is compelling them? Since the 1960s, all peoples of the South have been at least politically free. And since the early 1980s, the Pisanis and Narduccis and the EEC have even been telling them to be different. Nobody prevented us from using camels or bullock carts or wearing national dress. It is we who wanted to have cars and wear trousers.

Of course, there are many indirect (economic) compulsions. But, in principle, outsiders would have no objection if a people (or a part thereof) wished to revive, preserve and practise its traditional culture, as long as that did not harm others. But what do the peoples want? What are their dreams and visions? There is no doubt at all that the great majorities of all the peoples of the South want to catch up with what the North has achieved in the areas of economy, science, technology, education, and so on. In addition to their own traditional art, literature, music, they also want to learn and enjoy what the West has created in these areas. One only has to look at the Hindus, Buddhists, and Muslims of India, Thailand, and Egypt to be convinced about this. Whether this is good or not is a different question. No doubt, in social matters – for example, marriage and other ceremonies, rituals and customs in connection with birth and death – in religious matters, and in matters of clothing, most peoples of the South practise their traditional culture to some extent. But they do this only so far as it does not stand in the way of their efforts to catch up with the North in the areas of economy, science, technology and education.

Protagonists of the cultural approach want indigenous peoples to practise their traditional cultures. But what do they themselves want?

Let us take an example from India. Siddharta, a protagon of the cultural approach, writes about the Kurubas, a t Karnataka:

> Apart from imitating Hindu values they are also being infl ern and commercial values communicated through the cine youth do not know tribal songs, but will *enthusiastically* sing film songs in the Kannada language. … They still practise herbal medicine. But modern medicine has made deep inroads. For serious ailments many get admitted to government hospitals. … If the tribals are giving up some of their values, it is not entirely because of the aggression of the capitalist economy. To a certain extent *they themselves want these changes*. The space that the modern period creates for the individual is exhilarating to many tribal youth who find some of the customs of the clan oppressive (like the authority vested in the yajamana, or chief, for example). It must be admitted that the freedom of the individual and the space for that … is an important gain' (Siddharta 1991; emphasis added)

Similar things can be said about other tribal peoples of the world – about the North American Indians, and those living in the forests of Amazonia.

Industrial civilisation and capitalism

Verhelst writes: 'Like the Third World, the West is suffering from cultural uprooting'. That is also the opinion of Siddharta, who has lived for many years in France. This is confusing, for Western culture cannot be uprooted by Western culture. Who or what, then, is uprooting it? At this point, Verhelst suddenly makes other factors responsible for the evil, 'the great Promethean adventure … intensified during the last 300 years' and 'the *modern culture*' through which many Westerners have become 'egocentric' and 'creatures of domination and competitiveness' (all quotes in this paragraph are from Verhelst 1990: 72; emphasis added). With this analysis I fully agree. In plain English, these factors are industrial civilisation and capitalism, the destructive results of which the protagonists of the cultural approach are actually criticising but wrongly attributing to Western culture, which, according to Verhelst himself, 'contains much of great value'. The alternative programme should not, therefore, be a return to, or retrieval of, Eastern, traditional, regional or national cultures in opposition to Western culture, but eco-socialism with its rejection of both industrialism and capitalism.

For the eco-socialist movement, love of difference, tradition, and cultural identity is counterproductive. Also counterproductive is course to existing religions with the inglorious this-worldly history of their institutions and their advice to orient ourselves more towards the other world. In this connection, we should differentiate between religion, ethics, and spirituality. Like a baby, a religion demands total acceptance: one cannot accept only the good aspects and reject the bad ones. This is especially the case with religions based on allegedly divine revelations and holy scriptures. They are not amenable to discussion, debate, reason or logic. They cannot be open to new knowledge. Their opposition to, for example, artificial birth control or the dissolution of a marriage is honest. Their refusal to have dialogue with other religions is an indisputable consequence of faith.

We should not be suspicious of the slogan 'one world' simply because capitalist bosses also mouth it. In the history of humanity, all peoples have learned from one another: the Europeans took Christianity from the Palestinian Jews; Gandhi learned much from Tolstoy, Ruskin and Thoreau; many Europeans and Americans have learned from Gandhi and Lao Tse. The laws of nature are the same for all peoples, and the Earth has been one from the very beginning. The world, divided into many countries, must also become one for several practical purposes. The most important is to solve global ecological problems. But also to solve local ecological problems and the problem of hunger in poor countries, global co-operation and help from rich countries is necessary, at least for the next two or three decades. 'One world' does not necessarily mean the conception of the World Trade Organisation, the World Bank, the IMF and transnational corporations. Nor is anybody saying that Europeans and Americans must take the initiative in changing the traditional cultures of the South.

We must not forget that the peoples of the world are only so many variants of the same human species. The similarities between us are much greater than the differences, which are superficial. By similarities, I do not mean only our species-similarities; the peoples of the world are not only biologically similar; psychoanalysts have shown that our psyches, even our unconsciouses, are similar; we even dream similar dreams. Hitherto existing cultures are also largely similar. Institutions like marriage and family, belief in a God or in gods and goddesses, priesthood, temples, artistic expressions such as poetry, music and dance – all these things are almost universal. They existed among most peoples even before they had contact with each other. So, culturally, we are one humanity, despite minor differences. For all these reasons we must stick to the principle of one world, even though regional economies must become largely self-reliant.

Clash of civilisations?

We have been discussing whether to uphold or reject the ideal of one world and one humanity, but Samuel Huntington throws a bomb into the midst of our discussion with his hypothesis of the clash of civilisations (Huntington 1993a), which says that following the end of the Cold War, world politics is entering a new phase:

> the fundamental source of conflict in this new world will not be primarily ideological or primarily economic. The great divisions among humankind and the dominating source of conflict will be cultural. Nation states will remain the most powerful actors in world affairs, but the principal conflicts of global politics will occur between nations and groups of different civilizations. The clash of civilizations will dominate global politics. The fault lines between civilizations will be the battle lines of the future. (ibid: 22)

Elaborating his hypothesis, Huntington writes that he '... does not argue that civilization identities will replace all other identities, that nation states will disappear, that each civilization will become a single coherent political entity, that groups within a civilization will not conflict with and even fight each other' (ibid: 48). According to him, 'differences between civilizations are real and important; civilization-consciousness is increasing ...' (ibid). And, what is most important in our present context, he believes that

> cultural characteristics and differences are less mutable and hence less easily compromised and resolved than political and economic ones. In the former Soviet Union, communists can become democrats, the rich can become poor and the poor rich, but Russians cannot become Estonians and Azeris cannot become Armenians. In class and ideological conflicts, the key question was 'Which side are you on?' and people could and did choose sides and change sides. In conflicts between civilizations, the question is 'What are you?' That is a *given that cannot be changed*. And as we know, from Bosnia to the Caucasus to the Sudan, the wrong answer to that question can mean a bullet in the head. Even more than ethnicity, religion discriminates sharply and exclusively among people. A person can be half-French and half-Arab and simultaneously even a citizen of two countries. It is more difficult to be half-Catholic and half-Muslim. (ibid: 27; emphasis added)

Finally, for a better understanding of his views, it is necessary to know what for him is culture and what civilisation:

> A civilization is a cultural entity. Villages, regions, ethnic groups, nationalities, religious groups, all have distinct cultures at different levels of cultural heterogeneity. ... A civilization is ... the highest cultural grouping ... and

> the broadest level of cultural identity. ... People have levels of identity: a resident of Rome may define himself with varying degrees of intensity as a Roman, an Italian, a Catholic, a Christian, a European, a Westerner. The civilization to which he belongs is the broadest level of identification. (ibid: 23–4)

Huntington identifies seven or eight major civilisations in the present-day world: Western, Confucian, Japanese, Islamic, Hindu, Slavic-Orthodox, Latin American, and possibly African.

Critique of Huntington's hypothesis

After this short presentation of Huntington's hypothesis, I want to examine whether it is tenable. If what he says is right, then there is no hope of progress in human relations and relations between peoples. For in many countries, even in many towns and villages, people belonging to different cultures find themselves side by side. Huntington has referred to Bosnia and India. We can also point at the brutal conflicts in Lebanon between Muslims and Christians. And as to the civilisations, they too share humanity's one and only home, the planet Earth.

The array of facts and events with which Huntington supports his hypothesis is impressive, and at first sight convincing. In reply to his critics (Huntington 1993b), he calls his hypothesis the 'civilization paradigm' and claims that it is 'the best simple map of the post-Cold War world', and that it can account for many important developments in international affairs in recent years. He asks his critics, 'Can any other paradigm do better? If not civilizations, what?'

His critics – those who wrote in the journal *Foreign Affairs* (November–December 1993) – did not use the term 'paradigm'. But they tried to give other explanations for the facts and events of global politics in recent years and came to different conclusions about the future. After examining them, Huntington concluded that they provide no compelling alternative picture of the world. I agree.

One strand of alternative explanations is what Huntington calls 'the one-world paradigm'. It says that modernisation and economic development, the spread of the English language as *lingua franca*, increased interaction through wider communication and transport, have had a homogenising effect and produced a modern culture. According to this paradigm, a universal civilisation already exists or is likely to exist in the coming years.

Huntington rejects this one-world paradigm. I do too, but for different reasons. Huntington writes, recognising the fact of increased interactions, '... wars occur most frequently between societies with high levels of interaction, and interaction frequently reinforces existing identities and produces resistance, reaction and confrontation' (Huntington 1993b: 192). Moreover,

> A universal civilization can only be the product of universal power. Roman power created a near-universal civilization within the limited confines of the ancient world. ... European colonialism is over; American hegemony is receding. The erosion of Western culture follows, as indigenous, historically rooted mores, languages, beliefs and institutions reassert themselves. (ibid.)

I have three criticisms of these two statements. Firstly, it is not interaction *per se* that causes war, resistance, reaction and confrontation. These occur only if and when one party dominates, exploits or oppresses the other(s). That explains, for example, the Algerians' war of liberation against the French. Secondly, neither the Roman empire nor the European empires were one world. They were colonial worlds, which did not deserve to be called universal. Thirdly, even if indigenous cultures reassert themselves and existing identities are reinforced, why should that of itself lead to confrontation and war? Maybe (but I am not convinced) Russians cannot become Estonians, nor Azeris Armenians,[5] but why couldn't different cultures, identities and civilisations coexist side by side in peace, despite increased interaction?

Huntington cannot accept this position because he rejects the most important source of conflicts and wars in human history – conflicting economic interests and economic greed – as the primary source of current and future conflict. He himself suffers from the Single Alternative Fallacy, a criticism he levels against protagonists of the end-of-history theory. For if one does not want to admit that, for example, the war against Iraq (1991) was because of oil, then the only explanation that remains is the clash of civilisations.

I can offer an alternative paradigm for the limited purpose of explaining global politics. I believe it is much better than Huntington's. Let me call it the economic-interest paradigm. For reasons of space, I shall give only two examples. (1) The West helped Kuwaiti Muslims against Iraq because there is oil in Kuwait. It did not help Bosnian Muslims against the Serbs because there is no oil in Bosnia. (2) Egypt broke with the Soviet Union, which had helped it in war and peace, and became an ally of the USA, which had always helped its enemy, Israel. Why? Because the USA promised to be a better and more powerful *economic* ally than the Soviet Union. For the same reason, Egypt joined the American campaign against Iraq, an Islamic country.

The economic-interest paradigm is quite old, but it has not been so named. Referring to Thomas Kuhn, Huntington writes: 'A paradigm is disproved only by the creation of an alternative paradigm that accounts for more crucial facts in equally simple or simpler terms ...' (Huntington 1993b: 187). In this sense, the economic-interest paradigm has never been disproved. As Kuhn and Huntington say, some anomalous events do not falsify a paradigm. Huntington's civilisation paradigm is, therefore, not necessary, quite apart from being unconvincing.

Let us now take a few examples of conflict within a state: first, the break-up of the former Yugoslavia. According to Huntington, Croats and Slovenes belong to Western Christian civilisation, Serbs to the Slavic-Orthodox, and the Muslims of Bosnia to Islamic civilisation; and this was, for him, the reason for the break-up. But this does not explain why Croatia and Slovenia wanted and initiated the break-up, rather than Serbia, nor why the Muslims of Bosnia seceded only later. The economic-interest paradigm can explain these facts more easily. Croatia and Slovenia were the two most industrialised and otherwise prosperous (through tourism) republics, and they did not want to share their prosperity with the rest of Yugoslavia, which they had had to do in the 'socialist' federal state. This was also the reason why the Baltic rather than the Central Asian republics initiated the break-up of the USSR. And Russia, after it disavowed socialism, got rid of the Central Asian republics not because the majority of the people there were Muslims, but because these republics were, on the whole, economically uninteresting, even a burden.[6]

This paradigm can also explain why Turkey, despite the racist treatment Muslim Turks often receive in Europe, is trying to join the EU. It wants to join the club of the rich, and become rich with the EU's help. It can also explain why Christian Europeans had no problem importing Muslim labourers from the poor regions of Turkey, and why many in the rich northern provinces of Italy want to separate from the poorer southern provinces, although all are Catholic, white, and Italian-speaking.

Relative wealth and poverty provide the only convincing explanation why, in Germany, the Vietnamese are victims of xenophobia but not the Japanese, although, at least for Europeans, they look alike and belong to a different civilisation, and why Poles are hated and beaten up, but not, say, the French, although both are white and Western Christian. It is primarily the GNP of the state you come from and, secondarily, whether you have come to spend or earn money, that determines whether you are welcomed or hated in any European country.

It is similar in other parts of the world. In Indonesia, the growing chasm between rich and poor is generating ethnic and religious hatred and bloody attacks by the local, poor majority community against the rich or favoured immigrant minorities from other parts of the same country. Indonesian sociologist Adi Sasono cited: 'for example, the Chinese minority, which, in alliance with the authorities, controls two-thirds of the national prosperity. That this minority is also preponderantly Christian is in this context only of casual and secondary importance' (quoted in *Frankfurter Rundschau*, 13 January 1997).[7]

It is astonishing that Huntington ignores such overwhelming evidence. It is not as if the sources of conflict that he stresses do not exist. They do, but they are of secondary importance. Civilisational considerations are always subordinate to considerations of economic interest. The proof thereof is the fact that kin-state considerations referred to by Huntington play a role only if they do not conflict with a state's economic interests. The best example is Egypt's policy in America's war against Iraq.

Radical ecologists have long argued that conflict over resources would become a major source of war or civil war. In the past, the central government of Congo/Zaire waged war against the secessionist province, Shaba (Katanga), because it did not want to lose its mineral wealth. The central government of Nigeria waged war against the Ibos, who wanted to create their own state, Biafra, because they had oil in their province and did not want to share it with the other Nigerians. Morocco invaded and occupied Western Sahara (when Spain left) because the latter has large deposits of phosphate. And, as for the future, everyone, from the Worldwatch Institute to the World Bank, fears that in some regions of the world wars may break out because of disputes over fresh water – humans' most important resource.

The great tragedy of Rwanda (1994) was not a simple case of hatred generated through ethnic identity. Firstly, before 1959, when they were thrown out of power, the Tutsi minority, only 14 per cent of the population, had ruled the country. The hatred of the majority Hutus towards the Tutsis was therefore, at least partly, hatred of the oppressed and exploited towards the ruling aristocracy. And the contempt felt by the Tutsis towards the Hutus was that of the rulers towards the lower classes. Secondly, the rapid growth in the population – from about 3 million in 1962 to more than 7 million in 1993 – was the main, though not the only, cause of severe economic crisis in a small country with limited arable land and very modest industrial growth (cf. Dießenbacher 1994). The conflict could have been predicted, though not its severity.

So we leave the limited economic-interest paradigm and come back to the much broader and more general limits-to-growth paradigm. It is astonishing that both Huntington and his critics base their arguments on the assumptions of the growth paradigm, as if all peoples, nations, states, and civilisations could become rich, industrialised, developed, and modern. But if these assumptions are valid, then Huntington is wrong. Why should rich, developed, modern political entities fight each other? If their cultures are very different and if that is a problem, then they can live in a sort of cultural apartheid, in spite of necessary and growing interaction in the sphere of the economy. I cannot imagine any reason why economically equal and satisfied political entities should make war against each other.

But since the assumptions of the growth paradigm are wrong, the conclusion that the world has or soon would become one through industrialisation and modernisation is also wrong. One world is a vision, not already a reality. It is a long-term task, and a very difficult one. For it has to be realised on the basis of a low-level steady-state world economy, and tribal, ethnic, national, cultural, and civilisational loyalties and identities have to be defused.

We should thank Huntington for drawing our attention to some dangers and difficulties on our path towards one world. But surely he contradicts himself and supports the economic-interest paradigm when he writes:

> Decisions made at the UN Security Council or in the International Monetary Fund that reflect the interests of the West are presented to the world as reflecting the desires of the world community. The very phrase 'the world community' has become the euphemistic collective noun (replacing 'the Free World') to give global legitimacy to actions reflecting the interests of the United States and other Western *powers*. Through the IMF and other international economic institutions, the West *promotes its economic interests* and *imposes* on other nations the economic policies it thinks appropriate. (Huntington 1993a: 39; emphasis added)

To *this* West belongs Japan – in Huntington's view, not a part of Western civilisation – while Portugal, for example, which is not a power, does not belong. In fact, G–7 is a better term for this West.

With these sentences Huntington has unmasked the West; he has pinpointed its hypocrisy. Elsewhere in his essay he writes, with reference to Muslims contrasting Western actions against Iraq with the West's failure to protect Bosnian Muslims against Serbs: 'A world of clashing civilisations … is inevitably a world of *double standards*: people apply one standard to their kin-countries and a different standard to others' (ibid: 36; emphasis added). The double standard is obvious. But here Huntington makes a mistake: neither the Kuwaitis nor the Bosnian Muslims are kin of the West in his sense. The economic-interest paradigm gives a better explanation of this behaviour.

This unmasking of the West is a help for our movement. It makes clear that Western talk about values – democracy, freedom, human rights, and so on – and the West's efforts to propagate them in the rest of the world is a camouflage. Similarly, talk about identities and differences – cultural, ethnic, national, civilisational – can also be a camouflage used by ruling élites to promote and defend their own interests. Eco-socialists must develop their own new values and concepts.

SCIENCE, TECHNOLOGY, AND NATURE

I have often heard the remark that in the kind of eco-socialism I am proposing, not only would the economy fall back to the middle ages, but so also would science and technology. Let me first remove a confusion. Science and technology, though related, are two different things. Science is, firstly, a way of knowing; secondly, it also means the knowledge accumulated by applying scientific methods, and which can be considered reliable. Technologies, however, are methods of doing things, such as producing goods and services, repairing things, disposing of waste, and so on. Genetics is a science, genetic engineering a technology.

In eco-alternative literature, one can find much critique of modern or high technology. There is no doubt that there would be little future in an eco-socialist society for technologies that require a lot of resources and that are too complex. Not that there would be no machines at all in such a society. The main point is quantity. I guess that such a society would try to preserve the know-how of modern technologies and, for this purpose, decide to build and use some machinery and plant belonging to each modern technology, as long and as far as possible. But for mass use, the emphasis would be on intermediate, labour-intensive technologies, for reasons discussed in chapter 6.

I think so far there is agreement among radicals in the eco-alternative movement. But, unfortunately, there are many who criticise even science and its methods. On this issue, I do not agree with them. We must differentiate between social and natural sciences. In the former, the objects of research – people and their groups and societies – present many difficulties, so that the results are often vitiated. For instance, when questioned by social scientists, people do not always tell the truth about their life or views. In the natural sciences, such difficulties are fewer. Inanimate objects and animals do not lie. Behaviour of individuals within animal species of course differ, but even then, the difficulties are less than in the case of humans. If the criticism were limited to the social sciences and to sciences such as ethology[8] and evolution biology, I might have some understanding. But their criticism is general, without differentiation.

Let us take two examples: the International Society for Ecology and Culture writes under the sub-heading 'science and the reductionist world view':

> The economic paradigm goes hand in hand with modern science and technology; together they form the driving force behind industrial society.
>
> Science gains its understanding of the world by isolating and studying small pieces out of the interconnected continuum of nature. ... However, the ability of scientists to predict the consequences of their actions is limited to narrow parameters implicit in the scientific method: scientific models are most successful when dealing with the relatively simple and the short term. When it comes to the infinite complexity and long-term time-frame of social systems or ecosystems, the limitations of science are particularly evident. (ISEC 1992: 5–6)

Vandana Shiva writes under the sub-heading 'The violence of reductionism':

> I characterise modern western patriarchy's special epistemological tradition of the 'scientific revolution' as 'reductionist' because it reduced the capacity of humans to know nature both by excluding other knowers and other ways of knowing, and it reduced the capacity of nature to creatively regenerate and renew itself by manipulating it as inert and fragmented matter. Reductionism ... sees all systems as made up of the same basic constituents, discrete, unrelated and atomistic, and it assumes that all basic processes are mechanical. (Shiva 1988: 22)

These words are not criticism but accusations. One can accuse people of being violent, but one cannot accuse a way of knowing. One can criticise a government's science policy, or the scientific establishment, or, more concretely, the scientists who, for example, built the atom bomb, but not the science of physics or nuclear physics. How can a way of knowing be the driving force behind industrial society? It is capitalism, with its market forces and profit motive and, later, 'socialism' with its goal of catching up with the West and developing productive forces. A way of knowing can only be right or wrong, reliable or unreliable, suitable or unsuitable. If we find a better way of knowing, scientists will sooner or later accept it. It would become a part of science. But the present scientific way of knowing is the only reliable way we have.

If the method of 'isolating and studying small pieces' or working with 'fragmented matter' is the main criticism against science, then the critics must say how else could we have come to know that air consists of several gases, that oxygen is the most vital among them, or that carbon dioxide is the gas chiefly responsible for the greenhouse effect. Without 'reductionist' science, our knowledge would have remained at the level of the 15th century.

Moreover, science works not only by isolating and fragmenting: meteorologists study the movement of air as a whole, its velocity and temperature; ethologists observe living animals or groups of living animals; ornithologists observe the annual migration of birds from Siberia to India and back to Siberia; geographers study whole areas and human habitats; geologists study the whole Earth, or large parts of it (only geo-chemists isolate and fragment matter). The scientists who found the causes of the tides did not fragment the seas and rivers. And the science of ecology is the final proof that science is more than simply reductionist. The annual reports on the state of individual countries' and the world's environment would not be possible without modern science. And Lovelock's Gaia hypothesis is one hundred per cent scientific theory. There is nothing mystical or esoteric about it.

If science 'sees all systems as made up of the same basic constituents' (Shiva), of the elements, it is because that is the conclusion to which scientists had to come after four centuries of rigorous research. Wishful thinking about nature is of use neither to humans nor to the rest of nature. Moreover, science does not say that atoms or natural systems are 'unrelated' (Shiva's accusation); scientists speak of molecules and chemical compounds, and of ecosystems.

We have seen that many eco-alternative activists and theorists romanticise traditional cultures. Many of them also romanticise what they call traditional, vernacular, or people's knowledge systems. Shiva calls them 'non-reductionist knowledge systems' (Shiva 1988: 88). I have also come across the term 'vernacular science'. They demand that such knowledge systems have equal validity. For instance, the South's 'approach to Western science and technology' must be 'selective'; 'whatever science and technology we learn from outside must submit to the superior claim of our indigenous wisdom' (Kappen 1994). 'Given these fundamental shortcomings, the status of science today is profoundly disturbing. ... Science has come to dominate all other systems of knowledge. Traditions of non-western cultures and the experience and intuition of individuals are accepted only to the extent that they can be verified by scientific observation' (ISEC 1992: 6).

Here we see total confusion. Firstly, Kappen and ISEC mix up science and technology. We may find it advisable to reject this or that technology, this or that application of scientifically acquired knowledge. But we cannot reject science, the only reliable way of knowing. If we do that, we may fall prey to astrologers, shamans, gurus, and swallow all kinds of superstition and a hundred contradictory 'truths', 'wisdoms', 'knowledges' on the same question, none of which would be reliable. Even if such a thing as intuition exists, only a few can have it, and we would never know who really has it and who is a cheat.

Secondly, the experience and knowledge accumulated by traditional, village, and forest communities, and even their ways of knowing, are not, of course, totally worthless. But to say this is one thing; to criticise science for accepting them only to the extent that they are verifiable is another. Should science accept anything and everything that individuals or traditional communities say? After all, everybody can make mistakes. Moreover, such experience and knowledge are limited to their own lives and immediate surroundings. With such knowledge and experience we cannot tackle today's problems. They can be useful only in certain circumstances and only to complement modern knowledge.

As for questions regarding the meaning of life, life after death, and so on, no system of knowledge, neither traditional nor modern, knows anything reliable. In this sphere, only guesses, pleasant and unpleasant, are possible. The situation can be summed up:

> Faced with ... (pseudo)scientific prepotence, it is not surprising that grassroots ecological movements turn sometimes against science. ... anti-scientific activists sometimes conflate science and technology. Rightly mistrustful of unfounded scientific advice, aware of the deplorable ecological and social consequences of some technologies, grassroots ecologists sometimes turn against science per se, and get lost in the mists of irrationality. ... (Alier 1994: 32–3)

Alier credits peasants and tribal people with 'a steadying understanding of nature', which the grassroots ecologists he criticises no longer possess. But his own sympathies lie with the slogan: 'Scientific knowledge and revolution' (ibid).

An eco-socialist society would not reject modern science. A scientific attitude of mind would be promoted, rather than an esoteric, traditional or religious one, even if such a society could no longer finance research in, say, nuclear or particle physics, or astronomy. Eco-socialism is scientific socialism for the 21st century.

Attitude to nature

In the eco-alternative movement, one often finds a romanticisation of nature alongside a romanticisation of traditional and indigenous cultures and knowledge systems. Nature is beautiful, but to say that is one thing, and to say that nature is our mother, is benign, or is not an enemy is another. Of what use is such talk? Any theory built on such conceptions would only cause more confusion.

Nature is neither our mother nor our father. Nature is nature. If we think of nature as our mother, then not only mining, but also ploughing a field or felling a tree inflicts injuries on her. Nature is neither only benign nor only inimical to us. From nature we get food and other useful things, but nature also causes suffering through devastating floods, catastrophic earthquakes, disease, injuries, old age, and death. Nature is not only beautiful, but also ugly: there is brutal killing in nature, and even cannibalism; bullfrogs eat bullfrogs. There are power struggles in nature: male chimpanzees fight for domination over the group. In many species, the males fight each other for the exclusive right to copulate with the females.

Goethe, who in his critique of Newton reproached experimental science for vivisecting nature, which it could not bring back to life, also knew about nature's destructive side, and thought that 'the elements are to be viewed as colossal opponents'. He finally had to adopt a constructivist approach to nature as represented in the dam-building project in *Faust* (Böhme 1992: 41–3).

We have to accept these facts about nature. We cannot change them. Many may be unhappy to read such prosaic lines, but it is the only truthful way to perceive nature. For poets, of course, we must make an exception.

The famous speech of the Amerindian Chief Seattle[9] is beautiful literature, but its content is only partly true. Humans, including American Indians, need beasts not only to overcome 'great loneliness of spirit', but also as food. Let us be honest. Nobody knows whether stones and trees have a soul or not. Not even Amerindians know; they only imagine it.

It is all right for poets to call on us to learn from nature, or to say that nature knows best. But others should not say so: it is too simplistic. Although we could indeed learn from nature to adapt ourselves to our environment, human progress must mean more than just being part of nature. Erich Fromm speaks of humans being existentially split: 'to be part of nature and to transcend it by virtue of his consciousness' (Fromm 1973: 132). Our nature contains both good and evil. We must rise above our nature, otherwise we cannot speak of progress.

IS MORAL PROGRESS POSSIBLE?

The concept of eco-socialism is based on the assumption that people are capable of overcoming self-interest and that moral progress is possible. Workers in the first world have no *interest* in supporting the demand that third world workers get the same wages as they. They support it either for *moral* reasons or not at all. As to future generations, one can say: after me the deluge; only a moral sense prevents us from thinking like that. Many scientists and thinkers have examined human nature and tried to answer the question as to whether humans are basically good or evil. It is beyond my capacity and the scope of this book to deal with this question in detail. I shall therefore limit myself to a few facts and views relevant to the assumption that moral progress is possible.

The wolf has a bad image in human society. But zoologists have observed that wolves have an almost insuperable inhibition against biting weaker wolves (Weizsäcker *et al.* 1995: 326). In spite of that, many believe with Hobbes that humans are as bad as wolves: 'man is the wolf of man'. Poor wolves! Margaret Thatcher, the former Conservative Prime Minister of Great Britain, said that there is no such thing as society, there are only individuals. Among primates, our nearest relatives in the animal world, scientists have found morals and ethics. The renowned primatologist Frans de Waal shows that in their behaviour pattern not only aggression, but also affection, community feeling, sentiments of kindness and so on play a central role. According to de Waal, the capacity to be good is a potential given in the genes, even those of primates, and it can be activated through certain learning processes. De Waal observed among a group of Rhesus monkeys signs of a capacity to differentiate between good and evil, between right and wrong. He observed among bonobos and chimpanzees that after a fight the victor sought reconciliation with the vanquished. In cases where the opponents were totally unwilling to make peace, others took them by hand and led them to each other (de Waal 1996).

Of course, such behaviour is also observable among humans. But can it be hoped that humans, who stand on a higher rung of the ladder of biological evolution, are capable of reaching higher moral levels?

Aggression and destructiveness among humans

Many people think that the answer must be no. They have experienced and observed that humans are not merely as bad as wolves, but worse. They also find scientific support for their view, for example in Konrad Lorenz's theory of aggression. According to Lorenz,[10] the intra-specific aggression found among animals is the result of an instinct also present in humans. But there is, in this matter, a difference between animals and humans. Among animals, this aggression has a positive function for the survival of the species: it ensures that the individuals of the species are well spaced out in the available habitat; it ensures the selection of better males; and it ensures a social ranking. Aggression has fulfilled this function better ever since, in the evolution process, deadly aggression was transformed into symbolic and ritual threatening.

But, according to Lorenz, 'the so-called evil' among animals became among humans a real evil, a pernicious aggressive instinct. How this transformation took place during the tens of thousands of years of the palaeolithic age, is explained by Lorenz as follows:

> When man had reached the stage of having weapons, clothing and social organization, so overcoming the dangers of starving, freezing, and being eaten by wild animals, and these dangers ceased to be the essential factors influencing selection, an evil intra-specific selection must have set in. The factor influencing selection was now the wars waged between hostile neighbouring tribes. These must have evolved into an extreme form of all those so-called 'warrior virtues' ... (Lorenz 1976: 34)

Lorenz thinks (as does Freud, with his death instinct) that the human instinct of aggression is fed by a continuously flowing energy, which accumulates in the neural centre related to this instinct. This results in the build-up of pressure in a, so to speak, 'hydraulic' system. For him, therefore, not only is it unhealthy to suppress outward aggression, but it is also very difficult, if not impossible, to keep it under control. When too much pressure has built up, there can be an explosion: the human can become cruel, murderous, even without external stimulus. But, generally, both humans and animals find stimuli, search for stimuli, or themselves produce stimuli. Lorenz gives the example of founding a political party, which generates stimuli but is not the real cause of aggression.

If what Lorenz believes were true, there would not be much hope for moral progress. But there are grounds to believe that his theory is not wholly correct, and that the human condition is not so bad. Erich Fromm[11] differs from both Lorenz and Freud. Firstly, he does not agree that aggression, which indeed exists in both animals and humans as an instinct, works spontaneously and grows continuously like sexuality. It must first be mobilised through certain stimuli. When such stimuli are not present, aggression does not occur. Secondly, Fromm calls this reactive aggression, common to animals and humans, 'defensive' or 'benign' aggression.

Benign aggression serves 'vital interests'. Among animals, these are the individual's own life, care for progeny, access to individuals of the opposite sex, access to sources of food, and finally, survival of the species. Among humans it is basically the same, but for them vital interest can mean much more, which is why they can be more aggressive than animals. Humans can react not only to immediate but also to foreseeable future danger. Humans can create symbols and values with which they identify themselves so much that a threat to them becomes a threat to one's vital interests. Humans can become too attached to 'idols', which, in phases of deficient psychic development, may become necessary for their psychic survival. When such an 'idol' is attacked, the 'idolater' may see it as an attack against his or her vital interests. Finally, through education, ideology, and brainwashing, all kinds of 'vital interests' can be suggested to humans. Fromm thinks that 'all these factors … are results of the hitherto existing social structures, which were and are still based on the principle of exploitation and violence …' (quoted in Funk 1978: 186).

Fromm writes: 'If human aggression were more or less at the same level as that of other mammals … human society would be rather peaceful and nonviolent. But this is not so. Man's history is a record of extraordinary destructiveness and cruelty, and human aggression, it seems, far surpasses that of man's animal ancestors …' (Fromm 1973: 185). According to him, this 'hyperaggression' is unique to humans. 'What is unique in man is that he can be driven by impulses to kill and to torture, and that he feels lust in doing so; he is the only animal that can be a killer and destroyer of his own species without any rational gain, either biological or economic' (ibid: 218). Fromm calls this 'sadistic–cruel destructiveness', also 'malignant' aggression. Unlike benign aggression, it is not an instinct. When it occurs, very exceptionally, among animals, it is because the environmental and social balance has been disturbed, under circumstances such as 'crowding'.

One could argue that humans are so much more destructive because they have created or given rise to 'crowding' or other such aggression-generating conditions, which have become common. According to Fromm, that is to a large extent correct. But humans can be sadistically cruel and destructive even without external stimuli like 'crowding'; they can even (as Lorenz also says) await, search for, or bring about situations in which they can perpetrate destructive actions.

Fromm does not explain this phenomenon by the instinct of aggression, but by character. 'In the process of assimilation and socialisation, every human must in some way or other "relate" ("orient") himself or herself. The specific form of this being related finds expression in one's character and is at the same time the expression of one's character. "These orientations, through which the individual relates himself/herself to the world, constitute the core of his/her character" …' (Funk

1978: 53). The different types of human destructiveness are differently oriented reactions to human needs. Such different types of reaction can be explained by factors that shape character structures, so that no destructive instinct unique to humans need be assumed. The specifically human types of sadistic–cruel destructiveness are pathological forms of reaction to human needs. Fromm's understanding of sadistic–cruel destructiveness (malignant aggression) allows one to believe in the possibility of overcoming it, or at least sharply reducing it, because it is not determined by instinct but by character. Character is acquired and formed by external factors, hence, unlike an instinct, it can be changed by changing the external factors, namely social, economic, and political structures.

Fromm has three arguments against Lorenz's war hypothesis. Firstly, in view of the fact that in war aggressive individuals are killed in large numbers, it is not plausible that hyper-aggression among humans is due to a positive selection through war. With the dead young warriors, the gene responsible for such aggression should gradually disappear, leading to negative selection. Secondly, on the basis of overwhelming empirical anthropological evidence, he shows (quoting Q. Wright) that 'the collectors, lower hunters and lower agriculturists are the least warlike. The higher hunters and higher agriculturists are more warlike, while the highest agriculturists and the pastors are the most warlike of all' (Fromm 1973: 149). This proves to Fromm that bellicosity is not a function of a human instinct which manifests itself in the most primitive form of human society, but a function of the development of civilisation. The anthropological data show that the more division of labour prevails in a society, the more bellicose it is; and societies with a class system are the most bellicose. Thirdly, Fromm argues, before the advent of civilisation, primitive hunters and gatherers could not have much economic motivation to wage war. For population growth was negligible, there was not much movable property that could be robbed, and slaves could produce no surplus. Conflicts over hunting grounds or water holes were probably settled without battle. The stronger group just gradually pushed the weaker away.

At the end of his analysis, Fromm asks himself the question whether he has hope for humanity's future. He does not want to use the expression 'I am optimistic', because it sounds indifferent, reserved. He uses instead the term 'faith'. But faith is for Fromm 'rational faith', 'which is based on the clear awareness of all relevant data, and not, like "irrational faith", an illusion based on our desires' (ibid: 436). His position

> is one of rational faith in man's capacity to extricate himself from what seems the fatal web of circumstances that he has created. It is the position of those ... radicals who have rational faith in man's capacity to avoid the ultimate catastrophe. This humanist radicalism ... postulates that fundamental changes are necessary, not only in our economic and political structure but also in our values, in our concept of man's aims, and in our personal conduct.
>
> To have faith means to dare to think the unthinkable, yet to act within the limits of the realistically possible; it is the paradoxical hope ...' (ibid: 438).

Lorenz is also hopeful, but he does not speak of the necessity of changing society. He also has faith, but it is faith in evolution. He speaks of the necessity of an inhibiting mechanism in us, which would prevent violent aggression, and protect not only our personal friends but all humans from our instinctive aggressiveness. From this we must derive the commandment that we should love all humans, which is, of course, an age-old commandment (Lorenz 1976: 258).

> Our reason is quite able to understand its necessity, as our feeling is able to appreciate its beauty, but ... We can feel the full, warm emotion of friendship and love only for individuals, and the utmost exertion of will power cannot alter this fact. But the great constructors [mutation and selection] can, and I believe they will. I believe in the power of human reason, as I believe in the power of natural selection. I believe that reason can and will exert a selection pressure in the right direction. I believe that this, in the not too distant future, will endow our descendants with the faculty of fulfilling the greatest and most beautiful of all commandments. (ibid)

That is a rather long-term project for evolution. So, for the present, Lorenz gives a few concrete suggestions for controlling aggression: abreacting aggression on substitutes (the process of catharsis, sublimation), sport, promoting friendship between people of different nations, promoting art and science, humour, increasing the number of identifications with values, and many more.

Fromm's analysis of human aggression and destructiveness is convincing, but it contains a problem. One's understanding of 'vital interests' may include not only access to food and other economic goods required to satisfy basic needs, but also access to resources (such as oil) needed for the industrial mode of production and consumption. Today, therefore, it is not enough to say that aggression for the purpose of defending one's (or a nation's) vital interests is benign. What is required today is self-sacrifice, or the sacrifice of part of what we may understand as our vital interests.

Fromm seems in the early 1970s, when he wrote of vital interests, to have been unaware of the limits to growth. In one text he writes that hitherto existing social structures had to be based on exploitation and violence because of 'the underdeveloped state of the productive forces' (quoted in Funk 1978: 186). In the second half of the 1970s, however, he was fully aware of these limits, as evidenced in his book *To Have or To Be*. This awareness led him to write that there is 'a second argument ... in favour of profound psychological changes in Man as an alternative to economic and ecological catastrophe' (Fromm 1979: 17). Referring to the view of Mesarovic and Pestel, the authors of the second report to the Club of Rome, that only 'fundamental changes in the values and attitudes of man ... such as a new ethic and a new attitude towards nature' could 'avoid major and ultimately global catastrophy' (quoted in Fromm 1979: 18), he writes: '... a *new society* is possible *only if*, in the process of developing it, a *new human being* also develops ...' (ibid: 18; emphasis partly his, partly mine). But notwithstanding this new economic and ecological awareness, his new, ideal society remains industrial. He thinks two major tasks of the constructors of his new society would be 'to solve the problem of how to continue the industrial mode of production without total centralisation', and 'to give up the goal of unlimited growth for selective growth ...' (ibid: 170).

Fromm writes: 'The function of the new society is to encourage the emergence of a new Man' (ibid: 167). The qualities that the character structure of his new man will possess are all great. But in our finite Earth with finite resources, an industrial society, even if it opts for only selective growth, cannot generate the new man of his liking. Such a society will continue to need and generate *homo oeconomicus*. Aggression is not only war and naked violence: there is also what Johan Galtung (1975) calls 'structural violence'. The élites rob their poor generally through economic and political structures, that is, without any direct violence, although they also use naked violence when the exploited and oppressed start resisting.

Once more the new man

So we return to the theme of the new man, but now in a different context. We have seen how and why the project of the new man failed in the Soviet Union (chapter 3). Is there a chance that it might succeed in the context of building up an eco-socialist society?

Here we have a theoretical as well as a practical problem. The early Bolsheviks thought that the new man would emerge as a *result* of the socialist revolution and the revolutionary changes they were making in society. This corresponded to the Marxist theory that material being determines consciousness. Marxists believe in a necessary connection between socialist property relations and moral progress. 'Moral progress thus depends on the dis-alienation of man which depends on the revolution in property relations and the end of the division of labour, which in turn depends on "the expropriation of the expropriators" ' (Sklair 1970: 50). Moreover, the immense development of productive forces was, according to Marx and the Marxists, necessarily leading humanity to a socialist revolution, which would expropriate the expropriators.

But today, as stated earlier, we think moral progress is a *precondition* of success in averting catastrophes. Development of productive forces would not lead to moral progress. On the contrary, moral progress is necessary to stop the development of productive forces. Which must come first? An eco-socialist society or the new man? It looks almost like a chicken-and-egg problem. Fromm does not say that the new man will have to come before the new society, nor that he can come only as a result of the new society. He expects the new man to develop 'in the process of developing' the new society. I think he is right.

But perhaps, just as there are natural limits to economic growth, there are also in human nature limits to moral growth. Höltschi & Rockstroh, two reformists, write:

> Material wishes, striving for power, egoism etc. belong *from the very beginning* to humans. This cannot be changed by a 'cultural revolution' ... Of course, the values of humans shift constantly ... but in the foreseeable future, even in an ideal social order, humans will hardly become fully altruistic, eco-conscious, and peaceful beings. (Höltschi & Rockstroh 1985: 85)

This is what all self-styled realists say. If it were true, then, of course, there would be little probability of moral progress. But perhaps it is not true. 'If a sick person has even the barest chance of survival, no responsible physician will say, "Let's give up the effort", or will use only palliatives. On the contrary, everything conceivable is done to save the sick person's life. Certainly, a sick society cannot expect anything less' (Fromm 1979: 192).

In the preceding chapters I have shown that eco-capitalism, eco-Keynesianism, and market socialism will not cure the patient – the Earth and humanity – that they are at best only palliatives, having only temporary effect, and that only eco-socialism is the right medicine. There is, of course, no guarantee that eco-socialist politics today would succeed in bringing about an eco-socialist society. But, as Peter von Oertzen said, 'The ability of humans to be egoistic makes socialism necessary, their ability to be generous makes socialism possible' (quoted in *Frankfurter Rundschau*, 6 September 1994).

That humans have the ability to be generous is not wishful thinking. We witness it in everyday life. Moreover, egoism, greed for wealth and material goods, greed for power and status have not belonged to humans from their origins. Fromm has collected anthropological evidence to prove the contrary. E.R. Service, whose book is the most comprehensive presentation of anthropological findings on primitive hunters and gatherers, writes:

> We are accustomed ... to think that human beings have a 'natural propensity to truck and barter', and that economic relations ... are characterised by ... 'maximising' the result of effort, by 'selling dear and buying cheap'. Primitive peoples do none of these things, however; in fact, most of the time it would seem that they do the opposite. They 'give things away', they admire generosity, they expect hospitality, they punish thrift as selfishness. (quoted in Fromm 1973: 138)

They share food (such as meat) as a matter of course, it is the receiver's right to have a share. There is no question of thanking the giver, because it is all based on reciprocity (ibid: 139). On the question of property, Service writes:

> In no primitive band is anyone denied access to the resources of nature – no individual owns these resources. ...
>
> The natural resources on which the bands depend are collective, or communal, property. ... Within the band, all families have equal rights to acquire these resources. ...
>
> The things that seem most like private property are those that are made and used by individual persons. Weapons, knives and scrapers, clothing, ornaments, amulets, and the like, are frequently regarded as private property. ... Inasmuch as the possession of such things is dictated by their use, they are functions of the division of labour rather than an ownership of the 'means of production'. ... it is ... impossible to find in ethnographic accounts a case of some person or persons who, through some accident, owned no weapons or clothing and could not borrow or receive such things from more fortunate kinsmen. (ibid: 140)

Social relations among the members of primitive societies of hunters and gatherers are characterised by the absence of domination.

> There is no peck-order based on physical dominance at all, nor is there any superior–inferior ordering based on other sources of power such as wealth, hereditary classes, military or political office. The only consistent supremacy of any kind is that of a person of greater age and wisdom who might lead a ceremony.

> Even when individuals possess greater status or prestige than others, the manifestation of the high status and the prerogatives are the opposite of ape-like dominance. Generosity and modesty are required ... and the rewards they receive are merely the love or attentiveness of others. ... in primitive human society greater strength must be used in the service of the community, and the person, to earn prestige, must literally sacrifice to do so, working harder for less food. (ibid: 140–41)

Of course, it may not be scientifically correct to project descriptions of the life of primitive hunters and gatherers of the 19th and 20th century on to hunters and gatherers who lived before the rise of any civilisation. But these descriptions at least show that egoism, greed for material goods, hunger for power, hierarchy, and so on are not innate in humans. These character traits have been formed only in the course of civilisation.

Let us now look to the future. We know that the conception of socialism for which an industrial society with abundant material goods was an essential foundation could not but fail to develop the new man. In the 1920s and the 1930s, the Bolsheviks had to give up this ideal in order to catch up with the West economically. Can it be imagined that an eco-socialist society moving in the opposite direction might be fertile ground for the emergence of the new man? A remark of Service on primitive societies is relevant in this regard. 'It seems that the most primitive human societies are at the same time the most egalitarian. This must be related to the fact that because of *rudimentary technology* this kind of society depends on cooperation more fully more of the time than any other' (ibid: 141; emphasis added). And the following remarks of Fromm are relevant:

> Prehistoric hunters and agriculturists *had no opportunity* to develop a passionate striving for property or envy of the 'haves', because there was no private property to hold on to and no important economic differences to cause envy. ... There was no basis for the formation of the desire to exploit other human beings. ... Finally, there was little incentive for the development of greed, since *production and consumption were stabilized* at a certain level. (ibid: 160; emphasis added)

The technologies of an eco-socialist society would not, of course, be rudimentary. But they would be labour-intensive and simple. Production and consumption would be stabilised at a low level compared to today. Private property would be very limited, and economic differences would be negligible. The new man would have a much better chance to emerge in such a society.

Of course, the virtues of primitive hunters, gatherers, and

agriculturists did not extend beyond their own clan or tribe. But in modern times, with our level of knowledge, information and consciousness, they can extend to all peoples of the world. And although the project of the new man failed in the USSR, we know from accounts by Sinyavsky and the Webbs that for a few years the ideal was partly realised.

Weizsäcker, Lovins and Lovins –protagonists of eco-capitalism – say that 'in reality, egoism is not at all an unchangeable characteristic feature of humans dictated by the genes'. They refer to the works of Alfie Kohn, who has shown, on the basis of psychological research on co-operation, that even in modern capitalist societies, 'co-operation based on mutual respect is much more motivating and satisfying than competition' (Weizsäcker *et al.* 1995: 326). And Kenneth E. Boulding wrote as early as in 1969 in his essay 'Economics as a Moral Science':

> There is a widespread feeling that trade is somehow dirty and that merchants are somewhat undesirable characters, and that especially the labor market is utterly despicable as constituting the application of the principle of prostitution to virtually all areas of human life. This sentiment is not something that economists can neglect. We have assumed all too easily in economics that because something paid off it was therefore automatically legitimate. Unfortunately, the dynamics of legitimacy are more complex than this. Frequently it is negative payoffs, that is, sacrifices, rather than positive payoffs, which establish legitimacy. (Boulding 1969: 10)

We have seen above that Lorenz places his hope in evolution through selection. In this connection, a few remarks by Sahlins on palaeolithic hunters are very interesting:

> In selective adaptation to the perils of the Stone Age, human society overcame or subordinated such primate propensities as selfishness, indiscriminate sexuality, dominance and brute competition. It substituted kinship and co-operation for conflict, placed solidarity over sex, morality over might. In its earliest days it accomplished the greatest reform in history, the overthrow of human primate nature, and thereby secured the evolutionary future of the species. (quoted in Fromm 1973: 136–7)

It is not clear whether Sahlins here means adaptation through genetic changes. I do not think so. For qualities that come into being through genetic change do not vanish as quickly as co-operation, solidarity, and morality vanished in the course of development of our civilisations. It was surely simple adaptation to external circumstances and necessities; and the knowledge and experience of this adaptation was passed on to future generations, until it was no longer useful.

Similarly, in the near future, humans will be compelled to adapt themselves to ecological and resource-related necessities arising now, for the sake of sheer survival. In order to be successful in this effort, unless they are prepared to accept solutions through ecological catastrophe, war, civil war, genocide, famine, epidemics, and so on, they will have to create new societies, which, I believe, will have to be of an eco-socialist type. And for this adaptation, some character traits of primitive humans will again be necessary, namely co-operation, solidarity, and morality. That would be the new man.

TOWARDS ECO-SOCIALIST CULTURES

Many ecologists are convinced that the roots of our problems lie in our cultures. Thus, referring to the great dust bowl catastrophe that overtook the American Great Plains in the 1920s and 1930s, Walter and Dorothy Schwarz write: '… awareness is the first priority of any ecology. And what follows awareness? … Even awareness is not enough to ensure action, unless there is a shift of cultural attitude' (Schwarz 1987: 142).

Some of the protagonists of the cultural approach cited above also recognise the need for change or reform, but this must, according to them, remain within the framework of traditional/indigenous culture. For Agarwal *et al.*, 'the return to one's own culture' is 'not a blind harking back to traditions' (Agarwal *et al.* 1987: 351). Sachs writes: 'The opposite of development is by no means stagnation. … From Gandhi's "swaraj" to Zapata's "ejidos", there are in every culture visions of change' (Sachs 1989). I think many radical changes are necessary to solve the problems we are confronted with. But a culture that wants to change itself radically must be prepared to cease to be the same culture. If, say, Americans, in 20 years from today, give up their private cars and travel only by tram, train, bus and bicycle, eat only vegetarian food and dismiss their armed forces – all necessary for ecological and humanitarian reasons – then it would no longer be 'American' culture, but a new one.

We must leave behind traditional cultures as well as modern industrial–capitalist culture, however much we might be in love with or used to them. We must create new cultures. All hitherto existing cultures have proved incapable of tackling the great crises mankind is confronting today, crises which these cultures have themselves generated through omission and commission.

Why should anyone be afraid of new cultures? In the history of humanity, many cultures have disappeared or evolved in history to become new cultures. None has been there since the year dot. Today, to give only one example, we are witnessing a 'cultural revolution' among the Tuaregs of north Africa: a nomadic people, many of them are becoming sedentary agriculturists, with all the cultural consequences. They did not want this transformation, but ecological and economic necessity – drought and population growth – compelled them to accept it (cf. Stührenberg & Maitre 1997: 41–5).

It need not be one single new culture for all humanity, although I know of no argument against it, except that cultural diversity is beautiful. We can have several new cultures in future. But whatever gods and goddesses people may worship, whatever and however their food, clothing, arts, laws, customs and rituals may be, the cultures to be created must accept some common categorical imperatives, unknown to hitherto existing cultures when they originated. The most important is the ecological imperative. Everybody must recognise the limits to both economic and population growth. A large number of children must not be considered as God's gift, nor must the desire for a private car or at least one son be respected.

Everybody must also accept the imperative of equality. Without equality there will be no peace between societies, nor between men and women or rich and poor within societies. All the ideals of the Enlightenment and the French Revolution failed to bring peace because the principle of equality was reduced to mere equality before the law. The West's betrayal of these ideals is the original cause of today's counter-Enlightenment – religious fundamentalism, racism, nationalism, ethnic cleansing, xenophobia, and so on. It is mainly under conditions of exploitation, oppression, discrimination or contempt by another people, which is somehow dominating or superior, that an excessive, morbid, separatist need for identity arises. It may be called cultural identity, but it can be based on nation, tribe, race, clan, caste, religion or language. But a return to the genuine ideals of the Enlightenment is no longer adequate, for even the philosophers of the Enlightenment were unaware of ecological and demographic problems.

Another categorical imperative that new cultures must accept is that other species must be allowed to survive and have enough space – not only gorillas in Rwanda, lions in Gujrat (India), wolves in Europe, but also all the small animal and plant species. It is for this purpose, too, that we humans must reduce our numbers and, in the long run, withdraw from much of the territory we have occupied. Among all species, only human beings are capable of this conscious moral action, and to take it would be a sign of our moral progress. It is true that in the course of evolution many species have become extinct, but let us at least resolve that we ourselves will not cause the extinction of any more.

New cultures with such categorical imperatives can arise only in the framework of an eco-socialist economy and politics. But eco-socialism cannot remain only a matter of economy, politics and government in a narrow sense. It has to be understood as a new, comprehensive cultural framework. Progress would then mean advance towards such a cultural framework. An eco-socialist government would not come into existence until a movement to create such a cultural framework had first succeeded to some extent. And an eco-socialist government must, in turn, play a part in strengthening the movement, otherwise the government's stability could not be guaranteed.

I have no fear that our eco-socialist world, if it could be realised, might know only one monotonous culture. The future ecological and decentralised economies and the new regional institutions would provide enough space for different expressions of economic, social, political, religious, spiritual, literary and artistic life.

Change in values

Whatever we may call it – a shift in cultural attitude or the creation of new cultures – it must begin with a change in values. Some in the eco-alternative movement saw this change taking place as early as the second half of the 1970s. For example, Binswanger *et al.* saw 'a far-reaching change in our political consciousness and values', which, they hoped, 'will change our social relations and have effect on our economic system' (Binswanger *et al.* 1979: 214). Hartmut Bossel saw a change in values taking place, which he expected to 'exert strong social pressure' in the direction of an 'ecological alternative' (Bossel 1978: 19). We know today that these expectations have been disappointed. But why? Because the changes in values that were taking place then were misunderstood.

Ronald Inglehart, who made a thorough study of the phenomenon at the time, ascertained that '... the basic value priorities of Western publics had been shifting from a Materialist emphasis toward a Post-Materialist one – from giving top priority to physical sustenance and safety, toward heavier emphasis on belonging, self-expression and the quality of life' (quoted in Weinberger 1984: 13). His position was explained as follows:

> Basing himself on Maslow's theory of hierarchy of needs, Inglehart believes that ... *when material interests have been taken care of*, 'non-material' values like self-realisation, participation, aesthetic needs etc. come to the fore. Therefore, for the population group favoured with *middle-class origin* and

> education, the childhood and youth of which was moulded by *economic prosperity* of the post-war years, post-materialist values are of primary importance. (Brand 1982: 66; emphasis added)

This finding was right, but it was not a change in values, nor a 'silent revolution' (Inglehart 1977). Post-materialist values were only *added* to the list, once material interests had been abundantly taken care of, like cream on a cake. On the basis of prosperity, one demanded a better quality of life. Prosperity itself was not questioned or criticised, except by a very small minority. This explains why the radical ecological positions of the 1970s were so short-lived, and why the eco-alternative movement underwent a change of perspective in the 1980s and accommodated industrial prosperity in its programme of ecologically and socially restructuring industrial society (see Sarkar 1990).

Twenty years later, one observes a clear swing back to materialist values. Those who still enjoy prosperity are afraid of losing it, and the millions unemployed and underemployed are frantically looking for opportunities to earn money. A true change in values in the direction of an ecological and solidaristic alternative would be one that would not require today's prosperity as its material basis. An eco-socialist perspective would make that possible.

Religion or spirituality?

It is not enough to say that necessity will compel us to develop an eco-socialist society and the values and virtues of the new man. 'Necessity' may mean catastrophe, followed by enormous suffering. Can anything be done before catastrophes occur? In chapter 6, I have expressed my thoughts on eco-socialist politics for today and tomorrow. Can religion play a role in the movement for an eco-socialist culture? After all, religion has been an important element in all cultures.

Particularly in societies that have the industrial or financial capacity to prevent or tackle catastrophes for a few more decades, the question arises: what could the source of inspiration be to act *today*? There are many who believe that (only) religion can be that source. I shall not go into details, but I can give one good example of such an argument.

In an apparently well-argued paper, Catholic theologian Michael Kern (1996) tries to show what the special contribution of religion can be towards solving the ecological problem.

> To give good arguments (global justice, rights of future generations) is not difficult for ecological ethics ... but under which conditions are normative considerations translated into concrete action? ... here, religious value-

> orientations, which address deep-seated feelings and wishes, effect more than … pure social–ecological reason.

According to Kern, the 'death of God' has led to the 'death of nature'. Of course, to want to have ever more is a characteristic not exclusive to the modern age. But the disappearance of a religious understanding of the whole, which formerly gave security and meaning to life, is specifically modern. The modern attitude to life has been: 'if life is the last and only opportunity, then one must get everything out of it, experience everything, enjoy, consume'. And then comes what is known as compensatory consumption.

> Behind this striving for ever more … is a longing for wholeness and perfection, theologically speaking, a longing for paradise. It is, in the final analysis, … the 'troubled' heart wherein … the root of the ecological problematique is to be seen. For humans, because of their anthropological structure, this striving is indispensable. And yet, it cannot succeed. Humans are – that is the conclusion of philosophical anthropology – beings who need salvation. If they fail to recognize this *conditio humana*, they succumb to an arrogance, the consequences of which reach up to the ecological problematique. It is questionable whether at all 'homo compensator' can become 'homo oecologicus' without a religious bond with a transcendental instance which gives meaning to the whole world and to his/her own existence. (Kern 1996)

Since in the Christian faith salvation, God's grace, is given in advance as a gift, believers, according to Kern, 'enjoy an existential feeling of having arrived and being one with', which not only makes them happy but also gives them inner stability. Because of this feeling, believers do not have to search frantically for security, nor chase adventure; for them compensation becomes superfluous, and modest living becomes possible. Only such persons can have the strength to make the painful sacrifices necessary to solve the ecological crisis.

Of course, it would be wonderful if it functioned as Kern thinks. Unfortunately, it does not function like that. Proof lies in the fact that even before the modern age, when God was very much alive, even Popes, priests, and monks –first-rank believers, who had received God's grace in advance – felt so insecure or were so greedy that they even cheated other believers, through sales of indulgences, in order to amass wealth in their churches and cloisters, and live in luxury. Obviously, their 'paradise' was very much on this Earth. The fact that protestant ethics were so compatible with capitalism also disproves his thesis. It is, therefore, doubtful whether humans have a *need* for salvation, as Kern asserts.

It is not much different in the other religions. Hindu priests, Muslim mullahs, and Buddhist monks are not, generally, paragons of ecological virtue. Therefore, generally speaking, religions are not the source of the kind of inspiration we need to solve the ecological crisis. Of course, one may find examples that apparently prove the contrary – the simple lifestyle of Saint Francis of Assisi, Gandhi, some monks, mystics and ascetics – but they are exceptions that prove the rule. I do not think that present-day believing Christians who are committed eco-activists are so *because* they are Christians. They are good people who are also Christians: they would have this commitment, even if they had been born in an atheist family.

Kern's thesis also fails to explain a particular aspect of the economic behaviour of primitive hunters and gatherers, who had no conception of salvation or God's grace. E.R. Service writes: 'And strangest of all, the more dire the circumstances, the more scarce (or valuable) the goods, the less "economically" will they behave and the more generous do they seem to be' (quoted in Fromm 1973: 138). And Sahlins quotes from LeJeune's report (1887) on the Montagnais:

> In the famine through which we passed, if my host took two, three, or four Beavers, immediately, ... they had a feast *for all neighbouring Savages*. And if those people had captured something, they had one also at the same time; ... I told them that they did not manage well, and that it would be better to reserve these feasts for future days, ... They laughed at me. 'Tomorrow' (they said) 'we shall make another feast with what we shall capture'. Yes, but more often they capture only cold and wind. (Sahlins 1974: 30)

Of course, they did not manage well and deserved criticism. But the point here is that for their generosity, for their total lack of worry for the morrow, and for their lack of desire for accumulation, they did not need any faith in a transcendental instance in Kern's sense. They had faith in the solidarity and generosity of fellow savages. All they needed was their primitive communism. Modern people's feeling of insecurity, their need to provide for their individual future and the future of their children, and the resulting absence of solidarity, are due chiefly to our socio-economic system. Moreover, faith in God's grace and promise of salvation is a very weak protection against the compulsions of capitalism and competition, and against the constant bombardment of advertisement for more and ever more luxury goods.

The religion Kern has written about is no religion that we know. He himself often uses the term 'theology'. I have expressed my views on actually existing religions. Even if one could find something in them that tells their followers to treat nature gently, the objection remains that religions separate human groups from each other.

The transcendence that theologians such as Kern talk about comes close to spirituality. If we ignore the 'masters' and 'Gurus' who only want to make money through pseudo-spiritual talk, spirituality may indeed play a role in overcoming egoism and greed, and in generating solidarity and sympathy towards both fellow humans and the rest of nature. It can be humanistic without being anthropocentric. Indeed, it can be an antidote to the anthropocentrism of 'socialism' and other varieties of humanism.

In general, spirituality has one advantage. No variety of it, not even those that have arisen out of some religion, is as rigid, as exclusive, and as capable of dogmatism and intolerance as most religions. Fromm suggests a good possible synthesis between our anthropological need for spirituality (transcendence) and our basic scientific attitude.

> Indeed, for those who are not authentically rooted in theistic religion the crucial question is that of conversion to a humanistic 'religiosity' without religion, without dogmas and institutions, a 'religiosity' long prepared by the movement of non-theistic religiosity, from the Buddha to Marx. (Fromm 1979: 196–7)

It should perhaps be called non-theistic spirituality. When astronomers tell us about the vastness of the universe and physicists about sub-atomic particles, when biologists tell us about genes and Lovelock about Gaia, then we have material for wonder, meditation, deep thought about our place and short life-span in this universe, and our loneliness in it, all of which could motivate us to treat each other with compassion, love, solidarity and friendship.

The security of religious faith is lost for ever. There is no salvation. The death of God is final. It was bound to happen with the growth of scientific knowledge. A God, or gods and goddesses, for the existence of whom there is no evidence at all, cannot really be revived, much less give a meaning to our existence. That is part of our *conditio humana.*

So, if not religion, what can inspire us before catastrophes occur? In the history of humanity, there are many examples of a cause or an ideal inspiring people, including atheists. The early years of the Soviet Union and of communist China are two such examples. Today, eco-socialism may be able to inspire.

Height of a culture

The last point that needs to be discussed is the view of many critics that it is not only an eco-socialist economy that would be low-level, but also an eco-socialist culture, because of the low level of the economy. I have heard the view that a society whose economy is largely self-sufficient and hence largely dissociated from the world market would become like a frog in a well. Such critics mean culture not in the social-anthropological sense, but in the sense of art, literature, dance, music, and the products of intellect.

There is some truth in this. Although an eco-socialist society would emphasise cultural and intellectual progress instead of economic growth, there would be, compared to today, simply much less money for international exchanges of artists, musicians, writers, scientists, and so on. After all, air travel must also be drastically curtailed. An eco-socialist government might decide to finance such exchanges at the cost of other things, but the hundreds of thousands of air journeys to foreign countries undertaken every year by businessmen and tourists would no longer be possible. This need not mean, however, that travel to foreign countries would become impossible. Even when world trade and tourism had contracted drastically, there would be some foreign trade, for which sailors and ships would be needed. I can imagine that many young people would use this opportunity to see foreign countries. Eco-socialist governments might organise exchange programmes for teachers, workers, students, and so on. Some may decide to settle abroad.

But – and this is the main point – the general criteria for measuring the 'height' of a culture must change. Achievements in art, literature, music, science, philosophy should not be the essential criteria, but rather questions such as whether a society has abolished exploitation and oppression, whether it is exploiting other peoples, whether its members are free from hunger, whether the burdens of heavy and unpleasant work are distributed equally, whether patriarchy has vanished, whether the economic and socio-political organisation is such that no hierarchy is necessary for its proper functioning, and, above all, whether it is living in harmony with nature. Judged in terms of these criteria, all existing cultures are at a very low level.

CONCLUSION: EITHER ECO-SOCIALISM OR BARBARISM

Unlike the theory of Marxist socialism, the theory of eco-socialism – the scientific socialism for the 21st century – gives no guarantee of success. Although necessity may demand eco-socialist adaptation, human societies may fail to adapt. One society after another may break down; many are already breaking down and sinking in chaos and barbarism.

At the peak of the nuclear arms race, Edward Thompson (1980) wrote an essay called 'Exterminism, the Last Stage of Civilization'. I never believed that the United States and the Soviet Union would ever attack each other, so I never believed in the probability of nuclear extermination. In the meantime, however, another variety of exterminism has become reality. We call it genocide, an example of which we have recently seen in Rwanda.

Today, more than ever before, the alternatives are, in imitation of Rosa Luxemburg: either eco-socialism or barbarism.

NOTES

1. Dowry is an amount of money (and/or consumer durables) paid by the parents of a bride to the bridegroom or his parents. It is understood as the price for the 'favour' done by the latter to the former.

2. This and all following quotes from declarations of human and/or fundamental rights are from: Commichau (ed.) 1985.

3. For a more detailed presentation of my views on the topic, see Mies & Sarkar 1990.

4. The terms 'West' and 'North' mean the same thing in discussions of this kind. 'West' is generally used when the subject matter is culture, 'North' when it is development. I have used both because my subject matter is both culture and development.

5. Actually, Huntington contradicts himself. On the one hand, he says cultural or civilisational identity is 'a given that cannot be changed', and on the other hand, 'People can and do redefine their identities and, as a result, the composition and boundaries of civilizations change' (1993a: 24, 27). I agree with the second statement. The fact that a US identity originated in immigrants from many cultures and civilisations proves the point.

6. The anomaly that Russia was and is unwilling to let Chechenia leave the Russian Federation cannot be explained so simply. The explanation of this is the fear that if Chechenia becomes independent, then even the Russian Federation might break up.

7. The economic-interest paradigm would not suffice to explain the centuries-long persecution of the Jews in Europe. They were not very rich (except a few bankers and industrialists from the 19th century on) nor favoured, nor immigrants. I think the fact that their forefathers were responsible for the execution of Jesus Christ and the fact that they refused to become Christians also played a role.

8. Ethology is the science that studies the behaviour of animals and tries to come, on that basis, to conclusions about their nature.

9. Researchers have recently found out that the speech was actually written by a white American. But it does not matter who wrote it. The main point is that the sentiments expressed in it exist in the minds of some people.

10. The presentation of Lorenz's views is based on Lorenz 1976, Fromm 1973, and Funk 1978.

11. The presentation of Fromm's views is based on Fromm 1973 and Funk 1978.

Bibliography

(NB. The reader will see that in some cases I have quoted from or referred to the German version of a book although I have had access also to its English version and have mentioned both in the bibliography. This is because, in such cases, I had read the German version first. When I later tried to locate a particular quotation or information from the German version in the English one, I found in many cases that the latter had not only been abridged but also largely rewritten.)

Abalkin, Leonid and Anatoli Blinow (eds.) (1989) *Perestroika von Innen*, Düsseldorf.

Afanassjew, Juri (ed.) (1988) *Es gibt keine Alternative zu Perestroika*, Nördlingen.

Aganbegyan, Abel (1979) 'Exploiting Siberia's Natural Resources', in Goncharuk & Novikov (eds.) 1979.

Aganbegyan, Abel (1988) *The Challenge: Economics of Perestroika*, London.

Agarwal, Anil *et al.* (eds.) (1987) *The Fight for Survival*, New Delhi.

Agarwal, Anil and Sunita Naraian (1992) *Towards a Green World*, New Delhi.

Alier, Juan Martinez (1994) 'Ecological Economics and Ecosocialism', in Martin O'Connor (ed.) 1994.

Alt, Franz (1993) 'Heilung für den blauen Planeten – Schilfgras statt Atom', in *Wegweiser*, 4/1993.

Altner, Günter *et al.* (ed.) (1994) *Jahrbuch Ökologie 1994*, Munich.

Altvater, Elmar (1986) 'Alte Hüte mit grüner Feder', in Kallscheuer (ed.) 1986.

Altvater, Elmar (1990) 'Markt oder Plan – eine falsche Alternative', in Heine *et al.* (eds.) 1990.

Altvater, Elmar (1992) *Die Zukunft des Marktes*, Münster.

Altvater, Elmar (1993) *The Future of the Market*, London and New York.

Altvater, Elmar, Erika Hickel, Jürgen Hoffmann *et al.* (1986) *Markt, Mensch, Natur*, Hamburg.

Amalrik, Andrei (1980) *Can the Soviet Union Survive until 1984?*, Harmondsworth.

Amin, Samir (1977) 'Self-Reliance and the New International Economic Order', in *Monthly Review* Vol. 29. No. 3.

Antillan, S. (1996) 'Öko-Ausbeutung in Chiapas', in *ILA – Zeitschrift der Informationsstelle Lateinamerika*, No. 195, May.

Arbeitskreis Alternativenergie Tübingen (1982) *Energiepolitik von unten*, Frankfurt.

Bahro, Rudolf (1981) *The Alternative in Eastern Europe*, London.

Bahro, Rudolf (1987) *Logik der Rettung*, Stuttgart.

Bandyopadhyay, Jayanta and Vandana Shiva (1989) 'Political Economy of Ecology Movements', in *ifda dossier* 71, May/June.

Beckenbach, Frank *et al.* (ed.) (1985) *Grüne Wirtschaftspolitik – machbare Utopien*, Köln.

Beckerman, Wilfred (1972) 'Economists, Scientists and Environmental Catastrophes', in *Oxford Economic Papers*, November.

Beckerman, Wilfred (1995) *Small is Stupid*, London.

Bettelheim, Charles, Maurice Dobb *et al.* (1969) *Zur Kritik der Sowjetökonomie*, Berlin.

Bettelheim, Charles (1976) *Class Struggles in the USSR – First period*, New York, London.

Bettelheim, Charles (1978) *Class Struggles in the USSR – Second period*, Hassocks.

Bierter, Willy (1979) 'Der selbst-produzierende Konsument', in Brun (ed.) 1979.

Bildungswerk für Demokratie und Umweltschutz e.V. (1984) Annoucement for the congress,'Ökologie zwischen Selbstbeschränkung und Emanzipation', Berlin.

Binswanger, H.C., W. Geissberger and T. Ginsburg (1979) *Wege aus der Wohlstandsfalle*, Frankfurt.

Bischoff, Joachim and Michael Menard (1990) *Marktwirtschaft und Sozialismus – Der Dritte Weg*, Hamburg.

Block, J.R. and W. Maier (1984) *Wachstum der Grenzen*, Frankfurt.

Böge, Stefanie and Helmut Holzapfel (1994) 'Zu viele Kilometerfresser verderben den guten Geschmack', in *Frankfurter Rundschau*, 19 April.

Bogen, Ralf (1987) *Sowjetunion und 'Dritte Welt'*, Stuttgart.

Böhme, Hartmut (1992) 'Gaia – Views on Earth from Hesiod to James Lovelock', in Internationale Gesellschaft der bildenden Künste (ed.) 1992.

Bossel, Hartmut (1978) *Bürgerinitiativen entwerfen die Zukunft*, Frankfurt.

Boulding, Kenneth E. (1969) 'Economics as a Moral Science', in *The American Economic Review*, March.

Brakel, Manus van and Bertram Zagema (1994) *Sustainable Netherlands*, Amsterdam. (This is a summary of Maria Buitenkamp *et al.* 1992)

Brand, Karl-Werner (1982) *Neue Soziale Bewegungen*, Opladen.

Bress, Ludwig and Karl Paul Hensel (ed.) (1972) *Wirtschaftssysteme des Sozialismus im Experiment – Plan oder Markt?*, Frankfurt.

Brown, Lester R. (1994) 'Facing Food Insecurity', in Brown *et al.* 1994.

Brown, Lester R. *et al.* (1991) *State of the World 1991*, London.

Brown, Lester R. *et al.* (1994) *State of the World 1994*, New York and London.

Brown, Lester R. *et al.* (1995) *State of the World 1995*, New York and London.

Brun, Rudolf (ed.) (1979) *Der neue Konsument*, Frankfurt.

Buitenkamp, M., H. Venner and T. Wams (ed.) (1993) *Action Plan Sustainable Netherlands*, Amsterdam.

Buitenkamp, Maria *et al.* (Friends of the Earth Netherlands) (1992) *Sustainable Netherlands – Action Plan*, Amsterdam.

Capra, Fritjof (1996) Interview in *Frankfurter Rundschau*, 10 September.

Carr, E.H. (1971) *Foundations of a Planned Economy, Vol. 2*, London and Basingstoke.

Carr, E.H. (1976) *The Bolshevik Revolution 1917–1923 Vol. 2*, Harmondsworth.

Carr, E.H. and R.W. Davies (1974) *Foundations of a Planned Economy, Vol. 1*, Harmondsworth.

Commichau, Gerhard (ed.) (1985) *Die Entwicklung der Menschen – und Bürgerrechte von 1776 bis zur Gegenwart*, Göttingen, Zurich (all texts except one are in English).

Commoner, Barry (1971) *The Closing Circle – Nature, Man and Technology*, New York.

Commoner, Barry (1976) *The Poverty of Power – Energy and the Economic Crisis*, New York.

Conert, Hans Georg (1990) *Die Ökonomie des unmöglichen Sozialismus*, Münster.

Daly, Herman E. (1977) *Steady-State Economics*, San Francisco.

Daly, Herman E. (1990) 'Toward Some Operational Principles of Sustainable Development', in *Ecological Economics*, No. 2.

Daly, Herman E. (ed.) (1973) *Toward a Steady-State Economy*, San Francisco.

Daly, Herman E. (1973) 'Introduction', in Daly (ed.) 1973.

Daly, Herman E. (1980) 'Introduction to the Steady-State Economy', in Daly (ed.) 1980.

Daly, Herman E. (ed.) (1980) *Economics, Ecology, Ethics – Essays toward a steady-state economy*, New York and San Francisco.

Daly, Herman E. and John B. Cobb Jr (1990) *For the Common Good*, London.

Damus, Renate (1990) 'Jenseits von realem Sozialismus und Kapitalismus', in Heine *et al.* (ed.) 1990.

Deléage, Jean-Paul (1994) 'Eco-Marxist Critique of Political Economy', in O'Connor, Martin (ed.) 1994.

Der Spiegel (weekly journal) (various issues).

Devall, Bill and George Sessions (1985) *Deep Ecology – Living as if Nature Mattered*, Salt Lake City.

DGB (Deutscher Gewerkschaftsbund) (1985) *Umweltschutz und qualitatives Wachstum*, Düsseldorf.

Dhagamvar, Vasudha (1985) 'The Displaced, Their Past and Their Future', in *Lokayan Bulletin*, No. 3–4/5.

DIE (Deutsches Institut für Entwicklungspolitik) (1991) *Bericht vor dem Bundestagsausschuß für wirtschaftliche Zusammenarbeit über die Verschuldungsproblematik der Entwicklungsländer am 15.5.1991 in Berlin*, Berlin.

Die Grünen (1986) *Der Umbau der Industriegesellschaft*, Bonn.

Die Tageszeitung (daily newspaper) (various issues).

Dießenbacher, Hartmut (1994) 'Söhne ohne Land', in *Der Spiegel* No. 21.

DIW (Deutsches Institut für Wirtschaftsforschung) (1994) *Wirtschaftliche Auswirkungen einer ökologischen Steuerreform*, Berlin.

Djilas, Milovan (1958) *The New Class*, London.

Dobb, Maurice (1993) *Soviet Economic Development Since 1917*, New Delhi.

Dudde, Lasse (1996) 'Anderswo geht's', in *Green Peace Magazin* (German), July-August.

Edwards, Paul (ed.) (1967) *Encyclopaedia of Philosophy*, New York.

Ellenstein, Jean (1977) *Geschichte des 'Stalinismus'*, Berlin.

Encyclopaedia Americana (1982) International edition, Vol. 27, Danbury.

Engels, Friedrich (1972) 'On Authority', in Marx, Engels and Lenin 1972.
Federal Ministry of Environment (of Germany) (ed.) (1994) *Klimaschutz in Deutschland – Nationalbericht der Bundesregierung*, Bonn.
Ferenczi, Caspar and Brigitte Löhr (1987) *Aufbruch mit Gorbatschow?*, Frankfurt.
Fetscher, Iring (1980) *Überlebensbedingungen der Menschheit*, Munich.
Fischer, Joschka (1989) *Der Umbau der Industriegesellschaft*, Frankfurt.
Flavin, Christopher (1995) 'Harnessing the Sun and the Wind' in Brown *et al.* 1995.
Frankfurter Rundschau (daily newspaper) various issues.
French, Hilary F. (1991) 'Restoring the East European and Soviet Environments', in Brown *et al.* 1991.
Friedl, Christian (1995) 'Ökobilanz für Solaranlagen', in *Globus*, No. 9–10.
Fromm, Erich (1973) *The Anatomy of Human Destructiveness*, New York.
Fromm, Erich (1979) *To Have or To Be*, London.
Funk, Rainer (1978) *Mut zum Menschen*, Stuttgart.
Gabor, Dennis, Umberto Colombo, Alexander King and Riccardo Galli (1976) *Das Ende der Verschwendung – Zur materiellen Lage der Menschheit. Ein Tatsachenbericht an den Club of Rome*, Stuttgart. (The abridged English version of the book is entitled 'Beyond the Age of Waste' Oxford, 1978.)
Galtung, Johan (1975) *Strukturelle Gewalt – Beiträge zur Friedens – und Konfliktforschung*, Reinbek.
Gandhi, M. K. (1965) *My Varnashrama Dharma* (ed. Anand T. Hingorani), Bombay.
Gandhi, M. K. (1966) *Socialism of My Conception* (ed. Anand T. Hingorani), Bombay.
Gandhi, M. K. (1982) *Hind Swaraj*, Ahmedabad.
Gandhi, Mahatma (1997) 'The Quest for Simplicity – My Idea of Swaraj', in Rahnema and Bawtree 1997.
Georgescu-Roegen, Nicholas (1976) *Energy and Economic Myths*, New York.
Georgescu-Roegen, Nicholas (1986) 'The Entropy Law and the Economic Process in Retrospect', in *Eastern Economic Journal*, January-March.
Georgescu-Roegen, Nicholas (1992) Interview in *Greenpeace Magazin* (German) 1/1992.
Georgescu-Roegen, Nicholas (1978) 'Technology Assessment: The Case of the Direct Use of Solar Energy', in *Atlantic Economic Journal*, December.
Goldman, Marshall I. (1972) *The Spoils of Progress – Environmental Pollution in the Soviet Union*, Cambridge, Mass.
Goldman, Marshall I. (1983) *USSR in Crisis – The Failure of an Economic System*, New York and London.
Goldsmith, Edward, Martin Khor, Helena Norberg-Hodge, Vandana Shiva *et al.* (1992) *The Future of Progress – Reflections on Environment and Development*, Bristol and Berkeley.
Goncharuk, M. and V. Novikov (eds.) (1979) *Economic Growth and Resources*, Moscow.
Gorbachev, Mikhail (1987) *Perestroika*, New York.
Gorbunow, Eduard (1989) 'Plan oder Markt?' in Abalkin and Blinow (eds.) 1989.

Göricke, Fred v. and Monika Reimann (1982) 'Brasilien: das nationale Alkoholprogramm – eine verfehlte Energie-Investition', in Michelsen *et al.* (eds.) 1982.

Gorz, André (1983a) *Ecology as Politics*, London.

Gorz, André (1983b) *Farewell to the Working Class – an Essay on Post-Industrial Socialism*, London and Sydney.

Gorz, André (1985) *Paths to Paradise – On the Liberation from Work*, London and Sydney.

Grießhammer, Rainer (1996) 'Mehr virtuell als reell', in *Politische Ökologie* No. 49, November-December.

Hamel, H. (1972) 'Die Experimente der sozialistischen Marktwirtschaften', in Bress and Hensel (eds.) 1972.

Hardin, Garrett (1973) 'The Tragedy of the Commons', in Daly (ed.) 1973.

Hardin, Garrett (1980) 'Second Thoughts on "The Tragedy of the Commons"' in Daly (ed.) 1980.

Hayes, D. (1979) *Pollution – The Neglected Dimension* (Worldwatch Paper 27), Washington.

Heine, Michael *et al.* (eds.) (1990) *Die Zukunft der DDR-Wirtschaft*, Reinbek.

Heuler, Werner (1986) 'Sowjetische Energie – und Umweltpolitik vor und nach Tschernobyl', in *Kommune* No. 6/1986.

Heuser, Uwe Jean (1993) 'Geld, Freiheit, Ideologie', in *Zeit der Ökonomen* (a publication of *Die Zeit*), Hamburg.

Hickel, Rudolf (1986) 'Ökologisch-industrieller Komplex 2000', in Altvater, Hickel, Hoffmann *et al.* 1986.

Höltschi, René and Christian Rockstroh (1985) *Bausteine für Alternativen – Ota Siks Dritter Weg in ein Wirtschaftssystem der Nachmoderne*, Grüsch.

Huntington, Samuel P. (1993a) 'The Clash of Civilizations?', in *Foreign Affairs*, Summer.

Huntington, Samuel P. (1993b) 'If Not Civilizations, What?', in *Foreign Affairs*, November/December.

Inglehart, Ronald (1977) *The Silent Revolution*, Princeton.

Internationale Gesellschaft der bildenden Künste (ed.) (1992) *Terre Erde Tierra Earth*, Bonn.

IRB (Informationszentrum Raum und Bau der Frauenhofer-Gesellschaft) (ed.) (1985) *Biogasnutzung in der Landwirtschaft*, Stuttgart.

Irvine, Sandy (1995a) *Red Sails in the Sunset? – An ecopolitical critique of the socialist inheritance* (a publication of ECO, the Campaign for Political Ecology), Leeds.

Irvine, Sandy (1995b) *Sustainable Development – The Last Refuge of Humanism?*, Newcastle upon Tyne.

ISEC (International Society for Ecology and Culture) (1992) 'The Future of Progress', in Goldsmith *et al.* 1992.

Jablokow, Alexej (1988) 'Ökologische Ignoranz und ökologisches Abenteuertum' in Afanassjew (ed.) 1988.

Jacobs, Michael (1991) *The Green Economy*, London.

Kahn, Herman, William Brown and Leon Martel (1976) *The Next 200 Years – a Scenario for America and the World*, New York.
Kallscheuer, Otto (ed.) (1986) *Die Grünen – letzte Wahl?*, Berlin.
Kappen, S. (1994) 'Towards an Alternative Cultural Paradigm of Development', in *Lokayan Bulletin* No. 10/4.
Kern, Michael, (1996) 'Mißachtete Natur – glücksuchender Mensch', in *Politische Ökologie* 48, September/October.
Keynes, John Maynard (1931) *Essays in Persuasion*, London.
Khozin, G. (1976) *The Biosphere and Politics*, Moscow.
Klaus, Georg and Manfred Buhr (eds.) (1974) *Philosophisches Wörterbuch*, Vol. 1, (article on 'Arbeit'), Leipzig.
Klewe, Heinz (1996) 'RHAPIT, STORM und andere FRUITS', in *Politische Ökologie*, No. 49, November-December.
Kolakowski, Leszek (1978) *Main Currents of Marxism – Its Origin, Growth, and Dissolution Vol. 3*, Oxford.
Komarov, Boris (1980) *The Destruction of Nature in the Soviet Union*, White Plains and New York.
Kosta, Jiri, Jan Meyer, Sibylle Weber (1973) *Warenproduktion im Sozialismus*, Frankfurt.
Kothari, Smitu (1985) 'Ecology vs. Development', in *Lokayan Bulletin*, No. 3–4/5.
Krüger, Stephan (1990) 'Marktsozialismus – eine moderne Sozialismuskonzeption für entwickelte Länder', in Heine *et al.* (eds.) 1990.
Kuhn, Thomas. S (1962) *The Structure of Scientific Revolutions*, Chicago.
Kurz, Robert (1991) *Der Kollaps der Modernisierung – Vom Zusammenbruch des Kasernensozialismus zur Krise der Weltökonomie*, Frankfurt.
Laistner, Herman (1989) *Die Geduld der Erde geht zu Ende*, Frankfurt.
Lemeschew, Mikhail (1988) 'Wirtschaftsinteressen und soziale Naturnutzung', in Afanassjew (ed.) 1988.
Lenin, V.I. (1967a) *Selected Works, Vol. 2*, Moscow.
Lenin, V.I. (1967b) *Selected Works, Vol. 3*, Moscow.
Levinson, Charles (1980) *Vodka Cola*, Horsham.
Lohmann, Karl-Ernst (1985) 'Staatsplan versus Marktkonkurrenz', in Karl-Ernst Lohmann (ed.) *Sozialismus Passé?* (Argument-Sonderband AS 135), Berlin.
Lorenz, Konrad (1976) *On Aggression*, London.
Loske, Rainer *et al.* (1995) *Zukunftsfähiges Deutschland (Kurzfassung)*, Bonn.
Loske, Rainer *et al.* (1996) *Zukunftsfähiges Deutschland*, Basel.
Lovelock, J.E. (1987) *Gaia – A new look at life on Earth*, Oxford and New York.
Lovelock, J.E. (1990) 'Hands up for the Gaia Hypothesis', in *Nature*, 8 March.
Luks, Fred (1997) 'Der Himmel ist nicht die Grenze', in *Frankfurter Rundschau*, 21 January.
Madras Group (1983) 'What is Development? – Recalling an Old Debate', in *PPST Bulletin*, May.
Magdoff, Harry and Paul M. Sweezy (1990) 'Perestroika and the Future of Socialism', in Tabb (ed.) 1990.

Majumdar, R.C. *et al.* (1967) *An Advanced History of India*, London.
Malley, Jürgen (1996) 'Von Ressourcenschonung derzeit keine Spur', in *Politische Ökologie*, No. 49, November-December.
Mandelbaum, Kurt (1974) 'Sozialdemokratie und Imperialismus', in Kurt Mandelbaum, *Sozialdemokratie und Leninismus*, Berlin.
Marx, Karl (1954) *Capital*, Vol. 1, Moscow.
Marx, Karl (1981) *Capital*, Vol. 3, Harmondsworth.
Marx, Karl (1982) *Capital*, Vol. 1, Harmondsworth.
Marx, Karl and Friedrich Engels (1976) *Selected Works* (in 3 volumes) Vol. 3, Moscow.
Marx, Karl, Friedrich Engels and V.I. Lenin (1972) *Anarchism and Anarcho-Syndicalism*, Moscow.
Max-Neef, Manfred, Antonio Elizalde and Martin Hopenhayn (1990) *Human-Scale Development – An Option for the Future*, Motala.
McLaughlin, Andrew (1990) 'Ecology, Capitalism, and Socialism', in *Socialism and Democracy* (USA), No. 10, Spring/Summer.
McLaughlin, Andrew, (1993) *Regarding Nature – Industrialism and Deep Ecology*, Albany.
Meadows, Dennis *et al.* (1972) *The Limits To Growth*, London.
Meadows, Donella H., Dennis L. Meadows and Jorgen Randers (1992) *Beyond the Limits*, London.
Meek, Ronald L. (ed.) (1971) *Marx and Engels on the Population Bomb*, Berkeley.
Melíß, Michael *et al.* (1995) *Erneuerbare Energien – verstärkt nutzen*, Bonn.
Meyer-Abich, K.M. (1973) 'Die ökologische Grenze des herkömmlichen Wirtschaftswachstums', in Nussbaum (ed.) 1973.
Michelsen, Gerd and Öko-Institut Freiburg/Br. (ed.) (1991) *Der Fischer Öko-Almanach 91/92*, Frankfurt.
Michelsen, Gerd *et al.* (ed.) (1982) *Der Fischer Öko-almanach 82/83*, Frankfurt.
Mies, Maria and Saral Sarkar (1990) 'Menschenrechte und Bildung für alle?' in *Vorgänge*, October.
Moggridge, D.E. (1976) *Keynes*, Glasgow.
Müller, Eva (1993) *Das Ende der Ölzeit*, Frankfurt.
Mumford, Lewis (1970) *The Myth of the Machine*, New York.
Naess, Arne (1990) *Ecology, Community and Lifestyle*, Cambridge.
Naisbitt, John (1982) *Megatrends – Ten New Directions Transforming Our Lives*, New York.
Nick, Harry (1994) *Warum die DDR wirtschaftlich gescheitert ist*, Berlin.
Nove, Alec (1972a) '"Market Socialism" and its Critics', in *Soviet Studies*, July.
Nove, Alec (1972b) *An Economic History of the USSR*, Harmondsworth.
Nove, Alec (1975) 'Is There A Ruling Class in The USSR?', in *Soviet Studies*, Vol. 27, No. 4.
Nove, Alec (1982) *The Soviet Economic System*, London.
Nove, Alec (1983) *The Economics of Feasible Socialism*, London.
Nuber, Ursula (1994) 'Die produktive Kraft der Depression', in *Psychologie heute*, March.

Nussbaum, H. von (ed.) (1973) *Die Zukunft des Wachstums*, Düsseldorf.
O'Connor, James (1988) 'Capitalism, Nature, Socialism – A Theoretical Introduction', in *Capitalism Nature Socialism*, No. 1, Fall.
O'Connor, James (1994) 'Is Sustainable Capitalism Possible?', in O'Connor, Martin (ed.) 1994.
O'Connor, Martin (ed.) (1994) *Is Capitalism Sustainable? – Political Economy and the Politics of Ecology*, New York, London.
Omvedt, Gail and Govind Kelkar (1995) *Gender and Technology – Emerging Visions from Asia*, Bangkok (AIT).
Opschoor, Johannes B. (1991) *Environmental Taxes and Incentives* (a publication of Centre for Science and Environment), New Delhi.
Orton, David (1994) 'Struggling Against Sustainable Development – A Canadian Perspective', in *Z Papers*, January-March.
Orton, David (1995) *Some Limitations of a Left Critique and Deep Dilemmas in Environmental – First Nations Relationships* (*Green Web Bulletin*, No. 46), Saltsprings (Canada).
Palz, Wolfgang and Henri Zibetta (1991) 'Energy Pay-Back Time of Photovoltaic Modules', in *International Journal of Solar Energy*, Vol. 10.
Pattberg, Ursula (1992) 'Fallbeispiel Thailand – verfehlte Ressourcenpolitik', in *Informationsbrief (Sonderdienst) Weltwirtschaft & Entwicklung*, 29 June.
Pearce, David, Anil Markandya and Edward B. Barbier (1989) *Blueprint for a Green Economy*, London.
Pearce, Fred (1992) *The Dammed – Rivers, Dams, and the Coming World Water Crisis*, London.
Pepper, David (1993) *Eco-Socialism*, London.
Ponting, Clive (1991) *A Green History of the World*, London.
Postel, Sandra (1994) 'Carrying Capacity – Earth's Bottom Line', in Brown *et al.* 1994.
Rahnema, Majid and Victoria Bawtree (1997) *The Post-Development Reader*, London, New Jersey, Dhaka, Halifax and Cape Town.
Rheinisch-Westfälisches Institut für Wirtschaftsforschung (RWI) (1996) *Regionalwirtschaftliche Wirkungen von Steuern und Abgaben auf den Verbrauch von Energie – das Beispiel Nordrhein-Westfalen (Zusammenfassung)*, Essen.
Rosenbladt, Sabine (1986) *Der Osten ist grün*, Hamburg.
Rügemer, Werner (1992) 'Im Indianer-"Paradies" herrschten Diktatur und Kannibalismus', in *Frankfurter Rundschau*, 2 July.
Ryle, Martin (1988) *Ecology and Socialism*, London.
Sachs, Wolfgang (1989) 'Zur Archäologie der Entwicklungsidee' (an essay in 6 parts), in *EPD-EP* (various issues of the year).
Sachs, Wolfgang (1990) 'The Archaeology of the Development Idea', in *Interculture*, Vol. 23, No. 4, Fall.
Sadik, Nafis (UNFPA) (1990) *The State of World Population 1990*, New York.
Sahlins, Marshall (1974) *Stone Age Economics*, London.
Sarkar, Saral (1983) 'Marxism and Productive Forces – a Critique', in *Alternatives*, March.

Sarkar, Saral (1988) 'Von Gandhi lernen – aber wie?' in *Kommune*, 7/1988.
Sarkar, Saral (1990a) 'Accomodating Industrialism – a Third World View of the West German Ecological Movement', in *The Ecologist*, July/August.
Sarkar, Saral (1990b) 'Das Paradies von André Gorz – Fragen, Zweifel, Kritik', in *Kurswechsel* (Vienna), 3/1990.
Sarkar, Saral (1993a) *Green-Alternative Politics in West Germany Vol. 1 – The New Social Movements*, Tokyo and New Delhi.
Sarkar, Saral (1993b) 'Polemics is Useless – A proposal for an eco-socialist synthesis in the overpopulation dispute', in *Frontier* (Calcutta), 20 February and 13 March.
Sarkar, Saral (1994) *Green-Alternative Politics in West Germany, Vol. 2, The Greens*, Tokyo and New Delhi.
Saslawskaja, Tatjana (1989) *Die Gorbachowstrategie*, Vienna.
Schapiro, Leonard, Joseph Grodsom (eds.) (1981) *The Soviet Workers*, New York.
Schmidt-Bleek, Friedrich (1994) 'Ohne De-Materialisierung kein ökologischer Strukturwandel', in Altner *et al.* (eds.) 1994.
Schmitz-Schlang, Otmar (NABU) (1995) *Nachwachsende Rohstoffe – Chancen und Risiken*, Bonn.
Schneiders, Volker (1984) 'Die Ressourcen der Ökonomie', in Block and Maier 1984.
Schütte, Volker (1996) 'ArBYTE auf der Autobahn', in *Politische Ökologie*, No. 49, November-December.
Schwarz, Walter and Dorothy (1987) *Breaking Through: Theory and Practice of Wholistic Living*, Devon.
Sen, Mohit (1976) 'Did Gandhiji Have an Ideology?' in *New Age* (India), 18 April.
Senghaas, Dieter (1979) 'Dissoziation und autozentrierte Entwicklung – eine entwicklungspolitische Alternative für die Dritte Welt', in Senghaas (ed.) 1979.
Senghaas, Dieter (ed.) (1979) *Kapitalistische Weltökonomie – Kontroversen über ihren Ursprung und ihre Entwicklungsdynamic*, Frankfurt.
Sethi, Harsh and D.L. Sheth (1993) 'All Human Rights for All – A Report on the World Conference on Human Rights (NGO Forum) Vienna', in *PUCL Bulletin*, August, New Delhi.
Shiva, Vandana (1988) *Staying Alive – Women, ecology and survival in India*, New Delhi.
Siddharta (1991) 'Tribals in H.D. Kote Area – a case from India', in *Cultures and Development – quid pro quo*, No. 5–6.
Sik, Ota (1985) 'Dritter Weg und grüne Wirtschaftspolitik', in Beckenbach *et al.* (eds.) 1985.
Sik, Ota (1990) 'Die DDR zwischen Marktradikalität und Planungsdogmatismus', in Heine *et al.* (eds.) 1990.
Simon, Gabriela (1991) 'Wieviel ist zuviel?', in *blätter des Iz3W*, November.
Sinyavsky, Andrei (1990) *Soviet Civilization – a Cultural History*, New York.
Sivaraksa, Sulak (1996) 'Asiatische Identität aus einer buddhistischen Perspektive', in Wagner (ed.) 1996.

Sklair, Leslie (1970) *The Sociology of Progress*, London.
Smith, Hedrick (1976) *The Russians*, London.
Stalin, J.V. (1976) *Problems of Leninism*, Peking.
Stiftung Entwicklung und Frieden (1991) *Global Trends 1991*, Bonn.
Strahm, Rudolf H. (1981) *Überentwicklung – Unterentwicklung*, Gelnhausen.
Strong, Anna Louise (1956) *The Stalin Era*, Calcutta.
Strotmann, Peter, (1969) 'Preface' in Bettelheim *et al.* 1969.
Stührenberg, Michael and Pascal Maitre (1997) 'Die Herren der Wüste müssen seßhaft werden' in *Greenpeace Magazin*, No. 3.
Sünkens, Ralf (1995) *Krieg an der Küste*, Interview in *Spiegel Spezial* (Die neuen Energien) 7/1995.
Sweezy, Paul M. (1979) 'A Crisis in Marxian Theory', in *Monthly Review*, June.
Szymanski, Albert (1979) *Is the Red Flag Flying? – The Political Economy of the Soviet Union Today*, London.
Tabb, William K. (ed.) (1990) *The Future of Socialism – Perspectives from the Left*, New York.
The Ecologist (1992) 'Whose Common Future?' (special issue), July/August.
The Economist (weekly journal) (1985) 6 December.
Thompson, Edward (1980) 'Notes on Exterminism, the Last Stage of Civilization', in *New Left Review*, No. 121, London.
Tibbs, Hardin B.C. (1992) 'Industrial Ecology – An Environmental Agenda for Industry', in *Whole Earth Review*, Winter.
Trainer, F.E. (1985) *Abandon Affluence*, London.
Trainer, Ted (1996) 'Later Than You Think', in *Real World*, Autumn.
Tsuru, Shigeto and Helmut Weidner (eds.) (1985) *Ein Modell für uns – Die Erfolge der japanischen Umweltpolitik*, Köln.
Tüting, Ludmilla (1983) *Umarmt die Bäume – Die Chipko-Bewegung in Indien* (Sunderlal Bahuguna's account of the Chipko movement), Berlin.
Ullrich, Otto (1979) *Weltniveau – In der Sackgasse des Industriesystems*, Berlin.
Verhelst, T.G. (1990) *No Life Without Roots – Culture and Development*, London.
Waal, Frans de (1996) *Good Natured, The Origins of Right and Wrong in Humans and Other Animals*, Cambridge, Mass.
Wagner, Jost (ed.) (1996) *Thailands kritische Denker*, Trier.
Wahl, Peter (1997) 'Private Finanzströme – Königsweg oder Holzweg?' in WEED 1997.
WCED (World Commission on Environment and Development) (1987) *Our Common Future* (the 'Brundtland Report'), Oxford.
Webb, Sidney and Beatrice (1944) *Soviet Communism, A New Civilisation*, Calcutta and London.
WEED (World Economy Environment and Developoment) (1997) *Sculdenreport*, Bonn.
Weidner, Helmut (1985) 'Von Japan lernen? – Erfolge und Grenzen einer technokratischen Umweltpolitik', in Tsuru and Weidner (eds.) 1985.
Weinberger, Marie-Luise (1984) *Aufbruch zu neuen Ufern?*, Bonn.
Weissman, Steve (1971) 'Foreword' in Meek (ed.) 1971.
Weizsäcker, Ernst Ulrich von (1989) *Erdpolitik*, Darmstadt.

Weizsäcker, Ernst Ulrich von (1994) *Earth Politics*, London and New Jersey.
Weizsäcker, Ernst Ulrich von (1996) 'Die Ökosteuer ist kein Arbeitsplatzkiller', Interview in *Frankfurter Rundschau*, 13 June.
Weizsäcker, Ernst Ulrich von, Amory B. Lovins, and L. Hunter Lovins (1995) *Faktor Vier*, Munich.
Weizsäcker, Ernst von, Amory B. Lovins, and L. Hunter Lovins (1997) *Factor Four – Doubling Wealth, Halving Resource Use*, London.
Wicke, Lutz (1988) *Die ökologischen Milliarden*, Munich.
Wicke, Lutz, Lothar de Maizière, Thomas de Maizière (1990) *Öko-Soziale Marktwirtschaft für Ost und West*, München.
Wiles, Peter (1981) 'Wage and Income Policies', in Schapiro and Grodsom (eds.) 1981.

Index